M000283131

RESIDENTIAL WIRING

Based on the 2020 NEC®

2020
EDITION

JONES & BARTLETT
LEARNING

TABLE OF CONTENTS

TABLE OF CONTENTS (continued)

TABLE OF CONTENTS (continued)

TABLE OF CONTENTS (continued)

TABLE OF CONTENTS (continued)

TABLE OF CONTENTS (continued)

TABLE OF CONTENTS (continued)

TABLE OF CONTENTS (continued)

AMPACITIES OF INSULATED CONDUCTORS

Ampacities of Insulated Conductors with Not More Than Three Current-Carrying Conductors in Raceway, Cable, or Earth (Directly Buried)

Size AWG or kcmil	60°C (140° F) Types TW, UF	75°C (167° F) Types RHW, THHW, THW, THWN, XHHW, XHWN, USE, ZW	90°C (194° F) Types TBS, SA, SIS, FEP, FEPB, MI, PFA, RHH, RHW-2, THHN, THHW, THW-2.THWN-2, USE-2, XHH, XHHW, XHHW-2, XHWN-2, XHWN, XHHN, Z, ZW-2	60°C (140° F) Types TW, UF	75°C (167° F) Types RHW, THHW, THW, THWN, XHHW, USE	90°C (194° F) Types TBS, SA, SIS, THHN, THHW, THW-2, THWN-2, RHH, RHW-2, USE-2, XHH, XHHW, XHHW-2, XHWN, XHWN-2, XHHN	Size AWG or kcmil
	Copper			**Aluminum or Copper-Clad Aluminum**			
14*	15	20	25	–	–	-	
12*	20	25	30	15	20	25	12*
10*	30	35	40	25	30	35	10*
8	40	50	55	35	40	45	8
6	55	65	75	40	50	55	6
4	70	85	95	55	65	75	4
3	85	100	115	65	75	85	3
2	95	115	130	75	90	100	2
1	110	130	145	85	100	115	1
1/0	125	150	170	100	120	135	1/0
2/0	145	175	195	115	135	150	2/0
3/0	165	200	225	130	155	175	3/0
4/0	195	230	260	150	180	205	4/0
250	215	255	290	170	205	230	250
300	240	285	320	195	230	260	300
350	260	310	350	210	250	280	350
400	280	335	380	225	270	305	400
500	320	380	430	260	310	350	500
600	350	420	475	285	340	385	600
700	385	460	520	315	375	425	700
750	400	475	535	320	385	435	750
800	410	490	555	330	395	445	800
900	435	520	585	355	425	480	900
1000	455	545	615	375	445	500	1000
1250	495	590	665	405	485	545	1250
1500	525	625	705	435	520	585	1500
1750	545	650	735	455	545	615	1750
2000	555	665	750	470	560	630	2000

Notes:
1. Section 310.15(B) shall be referenced for ampacity correction factors where the ambient temperature is other than 30°C (86°F).
2. Section 310.15(C)(1) shall be referenced for more than three current-carrying conductors.
3. Section 310.16 shall be referenced for conditions of use.
*Section 240.4(D) shall be referenced for conductor overcurrent protection limitations, except as modified elsewhere in the Code.
See *Ugly's* page 3 for Adjustment Examples.
Source: NFPA 70®, *National Electrical Code*®. NFPA, Quincy, MA, 2020, Table 310.16, as modified.

AMPACITIES OF INSULATED CONDUCTORS

Ampacities of Single-Insulated Conductors in Free Air

Size	60°C (140° F)	75°C (167° F)	90°C (194° F)	60°C (140° F)	75°C (167° F)	90°C (194° F)	Size
AWG or kcmil	Types TW, UF	Types RHW, THHW, THW, THWN, XHHW, XHWN, USE, ZW	Types TBS, SA, SIS, FEP, FEPB, MI, PFA, RHH, RHW-2, THHN, THHW, THW-2,THWN-2, USE-2, XHH, XHHW, XHHW-2, XHWN, XHWN-2, XHHN, Z, ZW-2	Types TW, UF	Types RHW, THHW, THW, THWN, XHHW, USE	Types TBS, SA, SIS, THHN, THHW, THW-2, THWN-2, RHH, RHW-2, USE-2, XHH, XHHW, XHHW-2, XHWN, XHWN-2, XHHN	AWG or kcmil
	Copper			Aluminum or Copper-Clad Aluminum			
14*	25	30	35	-	-	-	-
12*	30	35	40	25	30	35	12*
10*	40	50	55	35	40	45	10*
8	60	70	80	45	55	60	8
6	80	95	105	60	75	85	6
4	105	125	140	80	100	115	4
3	120	145	165	95	115	130	3
2	140	170	190	110	135	150	2
1	165	195	220	130	155	175	1
1/0	195	230	260	150	180	205	1/0
2/0	225	265	300	175	210	235	2/0
3/0	260	310	350	200	240	270	3/0
4/0	300	360	405	235	280	315	4/0
250	340	405	455	265	315	355	250
300	375	445	500	290	350	395	300
350	420	505	570	330	395	445	350
400	455	545	615	355	425	480	400
500	515	620	700	405	485	545	500
600	575	690	780	455	540	615	600
700	630	755	850	500	595	670	700
750	655	785	885	515	620	700	750
800	680	815	920	535	645	725	800
900	730	870	980	580	700	790	900
1000	780	935	1055	625	750	845	1000
1250	890	1065	1200	710	855	965	1250
1500	980	1175	1325	795	950	1070	1500
1750	1070	1280	1445	875	1050	1185	1750
2000	1155	1385	1560	960	1150	1295	2000

Notes:
1. Section 310.15(B) shall be referenced for ampacity correction factors where the ambient temperature is other than 30°C (86°F).
2. Section 310.17 shall be referenced for conditions of use.
*Section 240.4(D) shall be referenced for conductor overcurrent protection limitations, except as modified elsewhere in the *Code*.
Source: NFPA 70®, *National Electrical Code*®, NFPA, Quincy, MA, 2020, Table 310.17, as modified.

AMPACITY CORRECTION AND ADJUSTMENT FACTORS

Examples

Ugly's *Residential Wiring* page 1 shows ampacity values for not more than three current-carrying conductors in a raceway or cable and an ambient (surrounding) temperature of 30°C (86°F).

Example 1: A raceway contains three 3 AWG THWN copper conductors for a three-phase circuit in an ambient temperature of **30°C.**

Ugly's *Residential Wiring* page 1, 75° column (under Copper) indicates **100 amperes.**

Example 2: A raceway contains three 3 AWG THWN copper conductors for a three-phase circuit at an ambient temperature of **40°C.** See page 1, 75°C column (under Copper), which indicates **100 amperes.** This value must be corrected because the ambient temperature is above 30°C. From NEC Table 310.15(B)(1), the ambient temperature correction factor for 40°C is **0.88.**

100 amperes x 0.88 = **88 amperes** = corrected ampacity

Example 3: A raceway contains six 3 AWG THWN copper conductors for two three-phase circuits at an ambient temperature of 30°C.

According to the table on page 1, the 75°C column (under Copper) indicates **100 amperes.** This value must be adjusted for more than three current-carrying conductors. The table on page 4 requires an adjustment of **80%** for four through six current-carrying conductors.

100 amperes x 80% = **80 amperes.** The adjusted ampacity is **80 amperes.**

Example 4: A raceway contains six 3 AWG THWN copper conductors for two three-phase circuits in an ambient temperature of 40°C. These conductors must be corrected for ambient temperature and adjusted for the number of current-carrying conductors

3

⚡ AMPACITY CORRECTION AND ADJUSTMENT FACTORS

According to the table on page 1, the 75° column (under Copper) indicates **100 amperes.**

According to NEC Table 310.15(B)(1), the 40°C ambient temperature correction factor is 0.**88.**

According to the table on page 4, the 4 through 6 conductor factor is 0.**80.**

100 amperes x 0.88 x 0.80 = **70.4 amperes.**

The new derated ampacity is **70 amperes.**

Adjustment Factors for More Than Three Current-Carrying Conductors

Number of Current-Carrying Conductors*	Percent of Values in Tables 310.16 Through 310.19 as Adjusted for Ambient Temperature if Necessary
4–6	80
7-9	70
10-20	50
21-30	45
31-40	40
41 and) above	35

*Number of conductors is the total number of conductors, spares included, in the raceway or cable adjusted in accordance with 310.15(E) and (F).
Source: NFPA 70®, *National Electrical Code*®, NFPA, Quincy, MA, 2016, Table 310.15(C)(1).

CONDUCTOR SIZING FOR SINGLE-PHASE DWELLING SERVICES AND FEEDERS

Single-Phase Dwelling Services and Feeders

Service or Feeder Rating (Amperes)	Conductor (AWG or kcmil)	
	Copper	Aluminum or Copper-Clad Aluminum
100	4	2
110	3	1
125	2	1/0
150	1	2/0
175	1/0	3/0
200	2/0	4/0
225	3/0	250
250	4/0	300
300	250	350
350	350	500
400	400	600

Note: This table is not permitted if adjustment or correction factors are required. Adapted from NEC® Table 310.12.

Table 310.12 can be used to size service conductors for a one-family dwelling where the service is single phase and is rated 100 through 400 amperes.

For two-family dwellings, multifamily dwellings, and three-phase electrical systems, see 310.12.

Table 310.12 can be used to size service conductors supplying the entire load associated with the one-family dwelling.

Table 310.12 can be used to size feeder conductors but those feeder conductors have to supply the entire load associated with the one-family dwelling.

Examples

What minimum size copper service conductors can be used to supply a 200-ampere, single-phase, 120/240-volt service for a one-family dwelling? These service conductors will supply the entire load associated with this dwelling.

CONDUCTOR SIZING FOR SINGLE-PHASE DWELLING SERVICES AND FEEDERS

In accordance with Table 310.12, 2/0 AWG copper conductors can be installed.

The service for a one-family dwelling will be a 400-ampere, single-phase, 120/240-volt service. The 400-ampere service will supply two 200-ampere feeders. Which minimum size copper conductors can be used to supply the 400-ampere service and which minimum size copper conductors can be used to supply each of the two 200-ampere feeders?

As the 400-ampere service conductors will be supplying the entire load associated with this one-family dwelling, it is permissible to use Table 310.12. Therefore, 400 kcmil copper conductors can be used to supply the 400-ampere service. The 200-ampere feeders are not supplying the entire load associated with this one-family dwelling, so it is not permissible to use Table 310.12. In accordance with Table 310.16, 3/0 AWG copper conductors have to be used to supply each of the two 200-ampere feeders.

 ## CONDUCTOR AND EQUIPMENT TEMPERATURE LIMITATIONS*

Examples

A 145-ampere noncontinuous load will be protected by a 150-ampere circuit breaker that is labeled for 75°C terminations. It would be permissible to use a 1/0 AWG THWN copper conductor that has a 75°C insulation rating and has an ampacity of 150 amperes (page 1).

When a THHN (90°C) conductor is connected to a 75°C termination, it is limited to the 75°C ampacity. Therefore, if a 1 AWG THHN copper conductor with a rating of 145 amperes were connected to a 75°C terminal, its ampacity would be limited to 130 amperes instead of 150 amperes. Therefore, a 1/0 AWG THHN copper conductor is required (page 1).

If the 145-ampere noncontinuous load listed above uses 1/0 THWN copper conductors rated for 150 amperes and the conductors are in an

* See *NEC* 110.14(C)(1)

CONDUCTOR AND EQUIPMENT TEMPERATURE LIMITATIONS*

ambient temperature of 40°C, the conductor's ampacity would have to be corrected for the ambient temperature.

According to /VFCTable 310.15(B)(1), the 40°C temperature correction factor = **0.88**

1/0 THWN = 150 amperes x 0.88 = **132 amperes**

A 1/0 THWN copper conductor has an ampacity of 132 amperes when the ambient temperature is 40°C.

As the load is 145 amperes, apply the ambient temperature correction factor to the next larger size THWN conductor.

2/0 THWN = 175 amperes (from the 75°C column—page 1).

175 amperes x 0.88 = **154 amperes.** This size is suitable for the 145-ampere load.

The advantage of using 90°C conductors is that you can apply ampacity derating factors to the higher 90°C ampacity rating, and it may save you from going to a larger conductor.

1/0 THHN = 170 amperes (from the 90°C column—page 1)

40°C temperature correction factor = 0.91 (90°C column—*NEC* Table 310.15(B)(1))

170 amperes x 0.91 = 154.7 = **154 amperes**

This size is suitable for the 145-ampere load.

This 90°C conductor can be used but can never have a final derated ampacity over the 1/0 THWN 75°C rating of 150 amperes.

You are allowed to use higher-temperature insulated conductors such as THHN (90°C) conductors on 60°C or 75°C terminals of circuit breakers and equipment, and you are allowed to correct and/or adjust from the higher value for temperature and number of conductors, but the final derated ampacity is limited to the 60°C or 75°C terminal insulation labels

* See *NEC* 110.14(C)(1)

VOLTAGE-DROP EXAMPLES

**Typical Voltage-Drop Values Based on Conductor Size
and One-Way Length (75°C Termination and Insulation)**

25 Feet									
		12 AWG	10 AWG	8 AWG	6 AWG	4 AWG	3 AWG	2 AWG	1 AWG
Amperes	20	1.98	1.24	0.78	0.49	0.31	0.25	0.19	0.15
	30		1.86	1.17	0.74	0.46	0.37	0.29	0.23
	40			1.56	0.98	0.62	0.49	0.39	0.31
	50				1.23	0.77	0.61	0.49	0.39
	60					0.92	0.74	0.58	0.46

50 Feet									
		12 AWG	10 AWG	8 AWG	6 AWG	4 AWG	3 AWG	2 AWG	1 AWG
Amperes	20	3.96	2.48	1.56	0.98	0.62	0.49	0.39	0.31
	30		3.72	2.33	1.47	0.92	0.74	0.58	0.46
	40			3.11	1.96	1.23	0.98	0.78	0.62
	50				2.46	1.54	1.23	0.97	0.77
	60					1.85	1.47	1.16	0.92

75 Feet									
		12 AWG	10 AWG	8 AWG	6 AWG	4 AWG	3 AWG	2 AWG	1 AWG
Amperes	20	5.94	3.72	2.33	1.47	0.92	0.74	0.58	0.46
	30		5.58	3.50	2.21	1.39	1.10	0.87	0.69
	40			4.67	2.95	1.85	1.47	1.16	0.92
	50				3.68	2.31	1.84	1.46	1.16
	60					2.77	2.21	1.75	1.39

🔌 VOLTAGE-DROP EXAMPLES

100 Feet									
		12 AWG	10 AWG	8 AWG	6 AWG	4 AWG	3 AWG	2 AWG	1 AWG
Amperes	20	7.92	4.96	3.11	1.96	1.23	0.98	0.78	0.62
	30		7.44	4.67	2.95	1.85	1.47	1.16	0.92
	40			6.22	3.93	2.46	1.96	1.55	1.23
	50				4.91	3.08	2.45	1.94	1.54
	60					3.70	2.94	2.33	1.85

125 Feet									
		12 AWG	10 AWG	8 AWG	6 AWG	4 AWG	3 AWG	2 AWG	1 AWG
Amperes	20	9.90	6.20	3.89	2.46	1.54	1.23	0.97	0.77
	30		9.30	5.84	3.68	2.31	1.84	1.46	1.16
	40			7.78	4.91	3.08	2.45	1.94	1.54
	50				6.14	3.85	3.06	2.43	1.93
	60					4.62	3.68	2.91	2.31

150 Feet									
		12 AWG	10 AWG	8 AWG	6 AWG	4 AWG	3 AWG	2 AWG	1 AWG
Amperes	20	11.88	7.44	4.67	2.95	1.85	1.47	1.16	0.92
	30		11.16	7.00	4.42	2.77	2.21	1.75	1.39
	40			9.34	5.89	3.70	2.94	2.33	1.85
	50				7.37	4.62	3.68	2.91	2.31
	60					5.54	4.41	3.49	2.77

A two-wire 20-ampere circuit using 12 AWG with a one-way distance of 25 feet will drop 1.98 volts.

120 volts – 1.98 volts = 118.02 = 118 volts at the load
240 volts – 1.98 volts = 238.02 = 238 volts at the load

9

🔌 CONDUCTOR PREFIXES

B—Braid (outer)

F—Fixture wire

FEP or FEPB—Fluorinated Ethylene Propylene (dry locations over 90°C)

H—(Heat) Operating temperatures up to 75°C (Lack of "H" indicates 60°C)

HH—Operating temperatures up to 90°C

L—Lead jacket (seamless)

MTW—Machine tool wire

N—Nylon jacket, resistant to oil and gas

O—Neoprene jacket

R—Thermoset insulation (examples are rubber and neoprene)

S—Appliance cord

SP—Rubber lamp cord

SPT—Plastic lamp cord

T—(Thermoplastic) Operating temperatures up to 60°C

U—Underground use

W—Moisture resistant

X—Cross-linked synthetic polymer insulation

🔌 AMPACITY OF CORDS—TYPES S, SJ, SJT, SP, ST

Flexible Cords

Wire Gauge	2 Conductor	3 Conductor	4 Conductor
18	10	7	6
16	13	10	8
14	18	15	12
12	25	20	16
10	30	25	20

MAXIMUM NUMBER OF CONDUCTORS IN ELECTRICAL METALLIC TUBING

Type Letters	Cond. Size AWG/kcmil	Trade Sizes in Inches									
		½	¾	1	1¼	1½	2	2½	3	3½	4
RHH, RHW, RHW-2	14	4	7	11	20	27	46	80	120	157	201
	12	3	6	9	17	23	38	66	100	131	167
	10	2	5	8	13	18	30	53	81	105	135
	8	1	2	4	7	9	16	28	42	55	70
	6	1	1	3	5	8	13	22	34	44	56
	4	1	1	2	4	6	10	17	26	34	44
	3	1	1	1	4	5	9	15	23	30	38
	2	1	1	1	3	4	7	13	20	26	33
	1	0	1	1	1	3	5	9	13	17	22
	1/0	0	1	1	1	2	4	7	11	15	19
	2/0	0	1	1	1	2	4	6	10	13	17
	3/0	0	0	1	1	1	3	5	8	11	14
	4/0	0	0	1	1	1	3	5	7	9	12
	250	0	0	0	1	1	1	3	5	7	9
	300	0	0	0	1	1	1	3	5	6	8
	350	0	0	0	0	1	1	3	4	6	7
	400	0	0	0	0	1	1	2	4	5	7
	500	0	0	0	0	1	1	2	3	4	6
	600	0	0	0	0	0	1	1	3	3	5
	700	0	0	u	0	0	1	1	2	3	4
	750	0	0	0	0	0	1	1	2	3	4
TW, THHW, THW, THW-2	14	8	15	25	43	58	96	168	254	332	424
	12	6	11	19	33	45	74	129	195	255	326
	10	5	8	14	24	33	55	96	145	190	243
	8	2	5	8	13	18	30	53	81	105	135
RHH*, RHW*, RHW-2*	14	6	10	16	28	39	64	112	169	221	282
	12	4	8	13	23	31	51	90	136	177	227
	10	3	6	10	18	24	40	70	106	138	177
	8	1	4	6	10	14	24	42	63	83	106
RHH*, RHW*, RHW-2*, TW, THW, THHW, THW-2	6	1	3	4	8	11	18	32	48	63	81
	4	1	1	3	6	8	13	24	36	47	60
	3	1	1	3	5	7	12	20	31	40	52
	2	1	1	2	4	6	10	17	26	34	44
	1	1	1	1	3	4	7	12	18	24	31
	1/0	0	1	1	2	3	6	10	16	20	26
	2/0	0	1	1	1	3	5	9	13	17	22
	3/0	0	1	1	1	2	4	7	11	15	19
	4/0	0	0	1	1	1	3	6	9	12	16
	250	0	0	1	1	1	3	5	7	10	13
	300	0	0	1	1	1	2	4	6	8	11
	350	0	0	0	1	1	1	4	6	7	10
	400	0	0	0	1	1	1	3	5	7	9
	500	0	0	0	1	1	1	3	4	6	7
	600	0	0	0	1	1	1	2	3	4	6
	700	0	0	0	1	1	1	1	3	4	5
	750	0	0	0	0	1	1	1	3	4	5
THHW, THWN, THWN-2	14	12	22	35	61	84	138	241	364	476	608
	12	9	16	26	45	61	101	176	266	347	443
	10	5	10	16	28	38	63	111	167	219	279
	8	3	6	9	16	22	36	64	96	126	161
	6	2	4	7	12	16	26	46	69	91	116
	4	1	2	4	7	10	16	28	43	56	71
	3	1	1	3	6	8	13	24	36	47	60
	2	1	1	3	5	7	11	20	30	40	51
	1	1	1	1	4	5	8	15	22	29	37

(continued on next page)

MAXIMUM NUMBER OF CONDUCTORS IN ELECTRICAL METALLIC TUBING

Type Letters	Cond. Size AWG/kcmil	Trade Sizes in Inches									
		½	¾	1	1¼	1½	2	2½	3	3½	4
THHN, THWN, THWN-2	1/0	1	1	1	3	4	7	12	19	25	32
	2/0	0	1	1	2	3	6	10	16	20	26
	3/0	0	1	1	1	3	5	8	13	17	22
	4/0	0	1	1	1	2	4	7	11	14	18
	250	0	0	1	1	1	3	6	9	11	15
	300	0	0	1	1	1	3	5	7	10	13
	350	0	0	1	1	1	2	4	6	9	11
	400	0	0	0	1	1	1	4	5	8	10
	500	0	0	0	1	1	1	3	5	6	8
	600	0	0	0	0	1	1	2	4	5	7
	700	0	0	0	0	1	1	2	3	4	6
	750	0	0	0	0	1	1	1	3	4	5
FEP, FEPB, PFA, PFAH, TFE	14	12	21	34	60	81	134	234	354	462	590
	12	9	1b	25	43	59	98	171	258	337	430
	10	6	11	18	31	42	70	122	185	241	309
	8	3	6	10	18	24	40	70	106	138	177
	6	2	4	7	12	17	28	50	75	98	126
	4	1	3	5	9	12	20	35	53	69	88
	3	1	2	4	7	10	16	29	44	57	73
	2	1	1	3	6	8	13	24	36	47	60
PFA, PFAH, TFE	1	1	1	2	4	6	9	16	25	33	42
PFA, PFAH, TFE,Z	1/0	1	1	1	3	5	8	14	21	27	35
	2/0	1	1	1	3	4	6	11	17	22	29
	3/0	u	1	1	2	3	5	9	14	18	24
	4/0	0	1	1	1	2	4	8	11	15	19
Z	14	14	25	41	72	98	161	282	426	556	711
	12	10	18	29	51	69	114	200	302	394	504
	10	6	11	18	31	42	70	122	185	241	309
	8	4	7	11	20	27	44	77	117	153	195
	6	3	5	8	14	19	31	54	82	10/	137
	4	1	3	5	9	13	21	37	56	74	94
	3	1	2	4	7	9	15	27	41	54	69
	2	1	1	3	6	8	13	22	34	45	57
	1	1	1	2	4	6	10	18	28	36	46
XHH, XHHW, XHHW-2, ZW	14	8	15	25	43	58	96	168	254	332	424
	12.	6	11	19	33	45	74	129	195	255	326
	10	5	8	14	24	33	55	96	145	190	243
	8	2	5	8	13	18	30	53	81	105	135
	6	1	3	6	10	14	22	39	60	78	100
	4	1	2	4	7	10	16	28	43	56	72
	3	1	1	3	6	8	14	24	36	48	61
	2	1	1	3	5	7	11	20	31	40	51
XHH, XHHW, XHHW-2	1	1	1	1	4	5	8	15	23	30	38
	1/0	1	1	1	3	4	7	13	19	25	32
	2/0	0	1	1	2	3	6	10	16	21	27
	3/0	0	1	1	1	3	5	9	13	17	22
	4/0	0	1	1	1	2	4	7	11	14	18
	250	0	0	1	1	1	3	6	9	12	15
	300	0	0	1	1	1	3	5	8	10	13
	350	0	0	1	1	1	2	4	7	9	11
	400	0	0	0	1	1	1	4	6	8	10
	500	0	0	0	1	1	1	3	5	6	8
	600	0	0	0	1	1	1	2	4	5	6
	700	0	0	0	0	1	1	2	3	4	6
	750	0	0	0	0	1	1	1	3	4	5

*Types RHH, RHW, and RHW-2 without outer covering.
See *Ugly's* page 175 for Trade Size/Metric Designator conversion.
Source: NFPA 70®, *National Electrical Code*®, NFPA, Quincy, MA, 2020, Table C.1, as modified.

MAXIMUM NUMBER OF CONDUCTORS IN RIGID METAL CONDUIT

Type Letters	Cond. Size AWG/kcmil	Trade Sizes in Inches											
		½	¾	1	1¼	1½	2	2½	3	3½	4	5	6
RHH, RHW, RHW-2	14	4	7	12	21	28	46	66	102	136	176	276	398
	12	3	6	10	17	23	38	55	85	113	146	229	330
	10	3	5	8	14	19	31	44	68	91	118	185	267
	8	1	2	4	7	10	16	23	36	48	61	97	139
	6	1	1	3	6	8	13	18	29	38	49	77	112
	4	1	1	2	4	6	10	14	22	30	38	60	87
	3	1	1	2	4	5	9	12	19	26	34	53	76
	2	1	1	1	3	4	7	11	17	23	29	46	66
	1	0	1	1	1	3	5	7	11	15	19	30	44
	1/0	0	1	1	1	2	4	6	10	13	17	26	38
	2/0	0	1	1	1	2	4	5	8	11	14	23	33
	3/0	0	0	1	1	1	3	4	7	10	12	20	28
	4/0	0	0	1	1	1	3	4	6	8	11	17	24
	250	0	0	0	1	1	1	3	4	6	8	13	18
	300	0	0	0	1	1	1	2	4	5	7	11	16
	350	0	0	0	1	1	1	2	4	5	6	10	15
	400	0	0	0	1	1	1	1	3	4	6	9	13
	500	0	0	0	1	1	1	1	3	4	5	8	11
	600	0	0	0	0	1	1	1	1	2	3	6	9
	700	0	0	0	0	1	1	1	1	3	3	6	8
	750	0	0	0	0	0	1	1	1	3	3	5	8
TW, THHW, THW, THW-2	14	9	16	25	44	59	98	140	215	288	370	581	839
	12	7	12	19	33	45	75	107	165	221	284	446	644
	10	5	9	14	25	34	56	80	123	164	212	332	480
	8	3	5	8	14	19	31	44	68	91	118	185	267
RHH*, RHW*, RHW-2*	14	6	10	17	29	39	65	93	143	191	246	387	558
	12	5	8	13	23	32	52	75	115	154	198	311	448
	10	3	6	10	18	26	41	58	90	120	154	242	350
	8	1	4	6	11	15	24	35	54	72	92	145	209
RHH*, RHW*, RHW-2*, TW, THW, THHW, THW-2	6	1	3	5	8	11	18	27	41	55	71	111	160
	4	1	1	3	6	8	14	20	31	41	53	83	120
	3	1	1	3	5	7	12	17	26	35	45	71	103
	2	1	1	2	4	6	10	14	22	30	38	60	87
	1	1	1	1	3	4	7	10	15	21	27	42	61
	1/0	0	1	1	2	3	6	8	13	18	23	36	52
	2/0	0	1	1	2	3	5	7	11	15	19	31	44
	3/0	0	1	1	1	2	4	6	9	13	16	26	37
	4/0	0	0	1	1	1	3	5	8	10	14	21	31
	250	0	0	1	1	1	3	4	6	8	11	17	25
	300	0	0	1	1	1	2	3	5	7	9	15	22
	350	0	0	0	1	1	1	3	5	6	8	13	19
	400	0	0	0	1	1	1	3	4	6	7	12	17
	500	0	0	0	1	1	1	2	3	5	6	10	14
	600	0	0	0	0	1	1	1	3	4	5	8	12
	700	0	0	0	0	1	1	1	2	3	4	7	10
	750	0	0	0	0	0	1	1	2	3	3	7	10
THHN, THWN, THWN-2	14	13	22	36	63	85	140	200	309	412	531	833	1202
	12	9	16	26	46	62	102	146	225	301	387	608	877
	10	6	10	17	29	39	64	92	142	189	244	383	552
	8	3	6	9	16	22	37	53	82	109	140	221	318
	6	2	4	7	12	16	27	38	59	79	101	159	230
	4	1	2	4	7	10	16	23	36	48	62	98	141
	3	1	1	3	6	8	14	20	31	41	53	83	120
	2	1	1	3	5	7	11	17	26	34	44	70	100
	1	1	1	1	4	5	8	12	19	25	33	51	74

(continued on next page)

MAXIMUM NUMBER OF CONDUCTORS IN RIGID METAL CONDUIT

Type Letters	Cond. Size AWG/kcmil	½	¾	1	1¼	1½	2	2½	3	3½	4	5	6
THHN, THWN, THWN-2	1/0	1	1	1	3	4	7	10	16	21	27	43	63
	2/0	0	1	1	2	3	6	8	13	18	23	36	52
	3/0	0	1	1	1	3	5	7	11	1b	19	30	43
	4/0	0	1	1	1	2	4	6	9	12	16	2b	36
	250	0	0	1	1	1	3	5	7	10	13	20	29
	300	0	0	0	1	1	3	4	6	8	11	1/	25
	350	0	0	1	1	1	2	3	5	7	10	15	22
	400	0	0	0	1	1	2	3	5	7	8	13	20
	500	0	0	0	1	1	1	2	4	5	7	11	16
	600	0	0	0	0	1	1	1	3	4	6	9	13
	700	0	0	0	0	1	1	1	1	3	5	8	11
	750	0	0	0	0	1	1	1	3	4	5	7	11
FEP, FEPB, PFA, PFAH, TFE	14	12	22	35	61	83	136	194	300	400	515	808	1166
	12	9	16	26	44	60	99	142	219	292	376	590	851
	10	6	11	18	32	43	71	102	157	209	269	423	610
	8	3	6	10	18	25	41	58	90	120	154	242	350
	6	2	4	7	13	17	29	41	64	8b	110	172	249
	4	1	3	5	9	12	20	29	44	59	77	120	174
	3	1	2	4	7	10	17	24	37	50	64	100	145
	2	1	1	3	6	8	14	20	31	41	53	83	120
PFA, PFAH, TFE	1	1	1	2	4	6	9	14	21	28	37	57	83
PFA, PFAH, TFE, Z	1/0	1	1	1	3	5	8	11	18	24	30	48	69
	2/0	1	1	1	3	4	6	9	14	19	2b	40	57
	3/0	0	1	1	1	2	5	8	12	16	21	33	47
	4/0	0	1	1	1	2	4	6	10	13	17	27	39
Z	14	15	26	42	73	100	164	234	361	482	621	974	1405
	12	10	18	30	52	71	116	166	256	342	440	691	997
	10	6	11	18	32	43	71	102	157	209	269	423	610
	8	4	7	11	20	27	45	64	99	132	170	267	386
	6	3	5	8	14	19	31	4b	69	93	120	188	271
	4	1	3	5	9	13	22	31	48	64	82	129	186
	3	1	2	4	7	9	16	22	35	47	60	94	136
	2	1	1	3	6	8	13	19	29	39	50	78	113
	1	1	1	2	5	6	10	15	23	31	40	63	92
XHH, XHHW, XHHW-2, ZW	14	9	15	25	44	59	98	140	215	288	370	581	839
	12	7	12	19	33	4b	75	107	165	221	284	446	644
	10	5	9	14	2b	34	56	80	123	164	212	332	480
	8	3	5	8	14	19	31	44	68	91	118	18b	267
	6	1	3	6	10	14	23	33	51	68	87	137	197
	4	1	2	4	7	10	16	24	37	49	63	99	143
	3	1	1	3	6	8	14	20	31	41	53	84	121
	2	1	1	3	5	7	12	17	26	35	45	70	101
XHH, XHHW, XHHW-2	1	1	1	1	4	5	9	12	19	26	33	52	76
	1/0	1	1	1	3	4	7	10	16	22	28	44	64
	2/0	0	1	1	2	3	6	9	13	18	23	37	53
	3/0	0	1	1	1	3	5	7	11	15	19	30	44
	4/0	0	1	1	1	2	4	6	9	12	16	25	36
	250	0	0	1	1	1	3	5	7	10	13	20	30
	300	0	0	1	1	1	3	4	6	9	11	18	25
	350	0	0	1	1	1	2	3	5	7	10	15	22
	400	0	0	0	1	1	2	3	5	7	9	14	20
	500	0	0	0	1	1	2	3	5	7	7	11	16
	600	0	0	0	0	1	1	1	3	4	5	9	13
	700	0	0	0	1	1	1	1	3	4	4	8	11
	750	0	0	0	0	1	1	1	3	4	5	7	11

*Types RHH, RHW, and RHW-2 without outer covering.
See Ugly's page 175 for Trade Size/Metric Designator conversion.
Source: NFPA 70®, *National Electrical Code*®, NFPA, Quincy, MA, 2020, Table C.9, as modified.

MAXIMUM NUMBER OF CONDUCTORS IN RIGID PVC CONDUIT, SCHEDULE 40

Type Letters	Cond. Size AWG/kcmil	½	¾	1	1¼	1½	2	2½	3	3½	4	5	6
RHH, RHW, RHW-2	14	4	7	11	20	27	45	64	99	133	171	269	390
	12	3	5	9	16	22	37	53	82	110	142	224	323
	10	2	4	7	13	18	30	43	66	89	115	181	261
	8	1	2	4	7	9	15	22	3b	46	60	94	137
	6	1	1	3	5	7	12	18	28	37	48	76	109
	4	1	1	2	4	6	10	14	22	29	37	59	85
	3	1	1	1	4	b	8	12	19	2b	33	52	75
	2	1	1	1	3	4	7	10	16	22	28	45	65
	1	0	1	1	1	3	5	7	11	14	19	29	43
	1/0	0	1	1	1	2	4	6	9	13	16	26	37
	2/0	0	0	1	1	1	3	5	8	11	14	22	32
	3/0	0	0	1	1	1	3	4	7	9	12	19	28
	4/0	0	0	1	1	1	2	4	6	8	10	16	24
	250	0	0	0	1	1	1	3	4	6	8	12	18
	300	0	0	0	1	1	1	2	4	5	7	11	16
	350	0	0	0	1	1	1	2	3	5	6	10	14
	400	0	0	0	1	1	1	1	3	4	6	9	13
	500	0	0	u	0	1	1	1	3	4	5	8	11
	600	0	0	0	0	1	1	1	2	3	4	6	9
	700	0	0	0	0	0	1	1	1	3	3	6	8
	750	0	0	0	0	0	1	1	1	2	3	5	8
TW, THHW, THW, THW-2	14	8	14	24	42	57	94	135	209	280	361	568	822
	12	6	11	18	32	44	72	103	160	21b	277	436	631
	10	4	8	13	24	32	54	77	119	160	206	325	470
	8	2	4	7	13	18	30	43	66	89	115	181	261
RHH*, RHW*, RHW-2*	14	5	9	16	28	38	63	90	139	186	240	378	546
	12	4	8	13	22	30	50	72	112	150	193	304	439
	10	3	6	10	17	24	39	56	87	117	150	237	343
	8	1	3	6	10	14	23	33	52	70	90	142	205
RHH*, RHW*, RHW-2*, TW, THW, THHW, THW-2	6	1	2	4	8	11	18	26	40	53	69	109	157
	4	1	1	3	6	8	13	19	30	40	51	81	117
	3	1	1	3	5	7	11	16	2b	34	44	69	100
	2	1	1	2	4	6	10	14	22	29	37	59	85
	1	0	1	1	3	4	7	10	1b	20	26	41	60
	1/0	0	1	1	2	3	6	8	13	17	22	35	51
	2/0	0	1	1	1	3	5	7	11	15	19	30	43
	3/0	0	1	1	1	2	4	6	9	12	16	25	36
	4/0	0	0	1	1	1	3	5	8	10	13	21	30
	250	0	0	1	1	1	3	4	6	8	11	17	25
	300	0	0	1	1	1	2	3	5	7	9	15	21
	350	0	0	0	1	1	1	3	5	6	8	13	19
	400	0	0	0	1	1	1	3	4	6	7	12	17
	500	0	0	0	1	1	1	2	3	5	6	10	14
	600	0	0	0	0	1	1	1	3	4	5	8	11
	700	0	0	0	0	1	1	1	2	3	4	7	10
	750	0	0	0	0	1	1	1	2	3	4	6	10
THHN, THWN, THWN-2	14	11	21	34	60	82	135	193	299	401	517	815	1178
	12	8	15	25	43	59	99	141	218	293	377	594	859
	10	5	9	15	27	37	62	89	137	184	238	374	541
	8	3	5	9	16	21	36	51	79	106	137	216	312
	6	1	4	6	11	15	26	37	57	77	99	156	225
	4	1	2	4	7	9	16	22	35	47	61	96	138
	3	1	1	3	6	8	13	19	30	40	51	81	117
	2	1	1	3	5	7	11	16	25	33	43	68	98
	1	1	1	1	3	5	8	12	18	25	32	50	73

(continued on next page)

MAXIMUM NUMBER OF CONDUCTORS IN RIGID PVC CONDUIT, SCHEDULE 40

Type Letters	Cond. Size AWG/kcmil	½	¾	1	1¼	1½	2	2½	3	3½	4	5	6
THHN, THWN, THWN-2	1/0	1	1	1	3	4	7	10	15	21	27	42	61
	2/0	0	1	1	2	3	6	8	13	17	22	35	51
	3/0	0	1	1	1	3	5	7	11	14	18	29	42
	4/0	0	1	1	1	2	4	6	9	12	15	24	35
	250	0	0	1	1	1	3	4	7	10	12	20	28
	300	0	0	1	1	1	3	4	6	8	11	17	24
	350	0	0	1	1	1	2	3	5	7	9	15	21
	400	0	0	0	1	1	1	3	5	6	8	13	19
	500	0	0	0	1	1	1	2	4	5	7	11	16
	600	0	0	0	1	1	1	1	3	4	5	9	13
	700	0	0	0	0	1	1	1	3	4	5	8	11
	750	0	0	0	0	1	1	1	2	3	4	7	11
FEP, FEPB, PFA, PFAH, TFE	14	11	20	33	58	79	131	188	290	389	502	790	1142
	12	8	15	24	42	58	96	137	212	284	366	577	834
	10	6	10	17	30	41	69	98	152	204	263	414	598
	8	3	6	10	17	24	39	56	85	110	150	237	343
	6	2	4	7	12	17	28	40	62	83	107	169	244
	4	1	3	5	8	12	19	28	43	58	75	118	170
	3	1	2	4	7	10	16	23	36	48	62	98	142
	2	1	1	3	6	8	13	19	30	40	51	81	117
PFA, PFAH, TFE	1	1	1	2	4	5	9	13	20	28	36	56	81
PFA, PFAH, TFE, Z	1/0	1	1	1	3	4	8	11	17	23	30	47	68
	2/0	0	1	1	3	4	6	9	14	19	24	39	56
	3/0	0	1	1	2	3	5	7	12	16	20	32	46
	4/0	0	1	1	1	2	4	6	9	13	16	26	38
Z	14	13	24	40	70	95	158	226	350	469	605	952	1376
	12	9	17	28	49	68	112	160	248	333	429	675	976
	10	6	10	17	30	41	69	98	152	204	263	414	598
	8	3	6	11	19	26	43	62	96	129	166	261	378
	6	2	4	7	13	18	30	43	67	90	116	184	265
	4	1	3	5	9	12	21	30	46	62	80	126	183
	3	1	2	4	6	9	15	22	34	45	58	92	133
	2	1	1	3	5	7	12	18	28	38	49	77	111
	1	1	1	2	4	6	10	14	23	30	39	62	90
XHH, XHHW, XHHW-2, ZW	14	8	14	24	42	57	94	135	209	280	361	568	822
	12	6	11	18	32	44	72	103	160	215	277	436	631
	10	4	8	13	24	32	54	77	119	160	206	325	470
	8	2	4	7	13	18	30	43	66	89	115	181	261
	6	1	3	5	10	13	22	32	49	66	85	134	193
	4	1	2	4	7	9	16	23	35	48	61	97	140
	3	1	1	3	6	8	13	19	30	40	52	82	118
	2	1	1	3	5	7	11	16	25	34	44	69	99
XHH, XHHW, XHHW-2	1	1	1	1	3	5	8	12	19	25	32	51	74
	1/0	1	1	1	3	4	7	10	16	21	27	43	62
	2/0	0	1	1	2	3	6	8	13	17	23	36	52
	3/0	0	1	1	1	3	5	7	11	14	19	30	43
	4/0	0	1	1	1	2	4	6	9	12	15	24	35
	250	0	0	1	1	1	3	5	7	10	13	20	29
	300	0	0	1	1	1	3	4	6	8	11	17	25
	350	0	0	1	1	1	2	3	5	7	9	15	22
	400	0	0	1	1	1	1	3	5	6	8	13	19
	500	0	0	0	1	1	1	2	4	5	7	11	16
	600	0	0	0	1	1	1	1	3	4	5	9	13
	700	0	0	0	1	1	1	1	3	4	5	8	11
	750	0	0	0	0	1	1	1	2	3	4	7	11

*Types RHH, RHW, and RHW-2 without outer covering.
See *Ugly's* page 175 for Trade Size/Metric Designator conversion.
Source: NFPA 70®, *National Electrical Code®*, NFPA, Quincy, MA, 2020, Table C.11, as modified.

DIMENSIONS OF INSULATED CONDUCTORS AND FIXTURE WIRES

Type	Size	Approx. Area Sq. In.	Type	Size	Approx. Area Sq. In.
RFH-2	18	0.0145	RHH*, RHW*, XF	10	0.0333
FFH-2	16	0.0172	RHW-2*, XFF		
RHW-2, RHH	14	0.0293	RHH*, RHW*, RHW-2*	8	0.0556
RHW	12	0.0353	TW, THW	6	0.0726
	10	0.0437	THHW	4	0.0973
	8	0.0835	THW-2	3	0.1134
	6	0.1041	RHH*	2	0.1333
	4	0.1333	RHW*	1	0.1901
	3	0.1521	RHW-2*	1/0	0.2223
	2	0.1750		2/0	0.2624
	1	0.2660		3/0	0.3117
	1/0	0.3039		4/0	0.3718
	2/0	0.3505		250	0.4596
	3/0	0.4072		300	0.5281
	4/0	0.4754		350	0.5958
	250	0.6291		400	0.6619
	300	0.7088		500	0.7901
	350	0.7870		600	0.9729
	400	0.8626		700	1.1010
	500	1.0082		750	1.1652
	600	1.2135		800	1.2272
	700	1.3561		900	1.3561
	750	1.4272		1000	1.4784
	800	1.4957		1250	1.8602
	900	1.6377		1500	2.1695
	1000	1.7719		1750	2.4773
	1250	2.3479		2000	2.7818
	1500	2.6938	TFN	18	0.0055
	1750	3.0357	TFFN	16	0.0072
	2000	3.3719	THHN	14	0.0097
SF-2, SFF-2	18	0.0115	THWN	12	0.0133
	16	0.0139	THWN-2	10	0.0211
	14	0.0172		8	0.0366
SF-1, SFF-1	18	0.0065		6	0.0507
RFH-1, TF, TFF, XF, XFF	18	0.0080		4	0.0824
TF, TFF, XF, XFF	16	0.0109		3	0.0973
TW, XF, XFF,	14	0.0139		2	0.1158
THHW, THW, THW-2				1	0.1562
TW, THHW,	12	0.0181		1/0	0.1855
THW, THW-2	10	0.0243		2/0	0.2223
	8	0.0437		3/0	0.2679
RHH*, RHW*, RHW-2*	14	0.0209		4/0	0.3237
RHH*, RHW*, RHW-2*,	12	0.0260		250	0.3970
XF, XFF				300	0.4608
				350	0.5242
				400	0.5863
				500	0.7073
				600	0.8676
				700	0.9887

(continued on next page)

DIMENSIONS OF INSULATED CONDUCTORS AND FIXTURE WIRES

Type	Size	Approx. Area Sq. In.	Type	Size	Approx. Area Sq. In.
THHN THWN THWN-2	750 800 900 1000	1.0496 1.1085 1.2311 1.3478	XHHW XHHW-2 XHH	300 350 400 500 600 700 750 800 900 1000 1250 1500 1750 2000	0.4536 0.5166 0.5782 0.6984 0.8709 0.9923 1.0532 1.1122 1.2351 1.3519 1.7180 2.0157 2.3127 2.6073
PF, PGFF, PGF, PFF, PTF, PAF, PTFF, PAFF	18 16	0.0058 0.0075			
PF, PGFF, PGF, PFF, PTF, PAF, PTFF, PAFF TFE, FEP, PFA FEPB, PFAH	14	0.0100			
TFE, FEP, PFA, FEPB, PFAH	12 10 8 6 4 3 2	0.0137 0.0191 0.0333 0.0468 0.0670 0.0804 0.0973	KF-2 KFF-2	18 16 14 12 10	0.0031 0.0044 0.0064 0.0093 0.0139
TFE, PFA, PFAH	1	0.1399	KF-1 KFF-1	18 16 14 12 10	0.0026 0.0037 0.0055 0.0083 0.0127
TFE, PFA, PFAH, Z	1/0 2/0 3/0 4/0	0.1676 0.2027 0.2463 0.3000			
ZF, ZFF, ZHF	18 16	0.0045 0.0061			
Z, ZF, ZFF, ZHF	14	0.0083			
Z	12 10 8 6 4 3 2 1	0.0117 0.0191 0.0302 0.0430 0.0625 0.0855 0.1029 0.1269			
XHHW, ZW XHHW-2 XHH	14 12 10 8 6 4 3 2	0.0139 0.0181 0.0243 0.0437 0.0590 0.0814 0.0962 0.1146			
XHHW XHHW-2 XHH	1 1/0 2/0 3/0 4/0 250	0.1534 0.1825 0.2190 0.2642 0.3197 0.3904			

*Types RHH, RHW, and RHW-2 without outer covenng
See *Ugly's* page 173 for conversion of square inches to mm².
Source: NFPA 70®, *National Electrical Code*®, NFPA Quincy, MA, 2020, Table 5, as modified.

18

DIMENSIONS AND PERCENT AREA OF CONDUIT AND TUBING

(For the combinations of wires permitted in Chapter 9, Table 1, *NEC*®)
(See *Ugly's* pages 173-174 for metric conversions.)

Trade Size Inches	Internal Diameter Inches	Total Area 100% Sq. Inches	2 Wires 31% Sq. Inches	Over 2 Wires 40% Sq. Inches	1 Wire 53% Sq. Inches	(NIPPLE) 60% Sq. Inches
Electrical Metallic Tubing (EMT)						
½	0.622	0.304	0.094	0.122	0.161	0.182
¾	0.824	0.533	0.165	0.213	0.283	0.320
1	1.049	0.864	0.268	0.346	0.458	0.519
1¼	1.380	1.496	0.464	0.598	0.793	0.897
1½	1.610	2.036	0.631	0.814	1.079	1.221
2	2.067	3.356	1.040	1.342	1.778	2.013
2½	2.731	5.858	1.816	2.343	3.105	3.515
3	3.356	8.846	2.742	3.538	4.688	5.307
3½	3.834	11.545	3.579	4.618	6.119	6.927
4	4.334	14.753	4.573	5.901	7.819	8.852
Electrical Nonmetallic Tubing (ENT)						
½	0.560	0.285	0.088	0.114	0.151	0.171
¾	0.760	0.508	0.157	0.203	0.269	0.305
1	1.000	0.832	0.258	0.333	0.441	0.499
1¼	1.340	1.453	0.450	0.581	0.770	0.872
1½	1.570	1.986	0.616	0.794	1.052	1.191
2	2.020	3.291	1.020	1.316	1.744	1.975
2½	–	–	–	–	–	–
3	–	–	–	–	–	–
3½	–	–	–	–	–	–
4	–	–	–	–	–	–
Flexible Metal Conduit (FMC)						
⅜	0.384	0.116	0.036	0.046	0.061	0.069
½	0.635	0.317	0.098	0.127	0.168	0.190
¾	0.824	0.533	0.165	0.213	0.283	0.320
1	1.020	0.817	0.253	0.327	0.433	0.490
1¼	1.275	1.277	0.396	0.511	0.677	0.766
1½	1.538	1.858	0.576	0.743	0.985	1.115
2	2.040	3.269	1.013	1.307	1.732	1.961
2½	2.500	4.909	1.522	1.963	2.602	2.945
3	3.000	7.069	2.191	2.827	3.746	4.241
3½	3.500	9.621	2.983	3.848	5.099	5.773
4	4.000	12.566	3.896	5.027	6.660	7.540
Intermediate Metal Conduit (IMC)						
⅜	–	–	–	–	–	–
½	0.660	0.342	0.106	0.137	0.181	0.205
¾	0.864	0.586	0.182	0.235	0.311	0.352
1	1.105	0.959	0.297	0.384	0.508	0.575
1¼	1.448	1.647	0.510	0.659	0.873	0.988
1½	1.683	2.225	0.690	0.890	1.179	1.335
2	2.150	3.630	1.125	1.452	1.924	2.178
2½	2.557	5.135	1.592	2.054	2.722	3.081
3	3.176	7.922	2.456	3.169	4.199	4.753
3½	3.671	10.584	3.281	4.234	5.610	6.351
4	4.166	13.631	4.226	5.452	7.224	8.179

(continued on next page)

DIMENSIONS AND PERCENT AREA OF CONDUIT AND TUBING

(For the combinations of wires permitted in Chapter 9, Table 1, *NEC®*)
(See *Ugly's* pages 173-174 for metric conversions.)

Trade Size Inches	Internal Diameter Inches	Total Area 100% Sq. Inches	2 Wires 31% Sq. Inches	Over 2 Wires 40% Sq. Inches	1 Wire 53% Sq. Inches	(NIPPLE) 60% Sq. Inches
Liquidtight Flexible Nonmetallic Conduit (Type LFNC-B*)						
⅜	0.494	0.192	0.059	0.077	0.102	0.115
½	0.632	0.314	0.097	0.125	0.166	0.188
¾	0.830	0.541	0.168	0.216	0.287	0.325
1	1.054	0.873	0.270	0.349	0.462	0.524
1¼	1.395	1.528	0.474	0.611	0.810	0.917
1½	1.588	1.981	0.614	0.792	1.050	1.188
2	2.033	3.246	1.006	1.298	1.720	1.948
*Corresponds to Section 356.2(2).						
Liquidtight Flexible Nonmetallic Conduit (Type LFNC-A*)						
⅜	0.495	0.192	0.060	0.077	0.102	0.115
½	0.630	0.312	0.097	0.125	0.165	0.187
¾	0.825	0.535	0.166	0.214	0.283	0.321
1	1.043	0.854	0.265	0.342	0.453	0.513
1¼	1.383	1.502	0.466	0.601	0.796	0.901
1½	1.603	2.018	0.626	0.807	1.070	1.211
2	2.063	3.343	1.036	1.337	1.772	2.006
*Corresponds to Section 356.2(1).						
Liquidtight Flexible Metal Conduit (LFMC)						
⅜	0.494	0.192	0.059	0.077	0.102	0.115
½	0.632	0.314	0.097	0.125	0.166	0.188
¾	0.830	0.541	0.168	0.216	0.287	0.325
1	1.054	0.873	0.270	0.349	0.462	0.524
1¼	1.395	1.528	0.474	0.611	0.810	0.917
1½	1.588	1.981	0.614	0.792	1.050	1.188
2	2.033	3.246	1.006	1.298	1.720	1.948
2½	2.493	4.881	1.513	1.953	2.587	2.929
3	3.085	7.475	2.317	2.990	3.962	4.485
3½	3.520	9.731	3.017	3.893	5.158	5.839
4	4.020	12.692	3.935	5.077	6.727	7.615
Rigid Metal Conduit (RMC)						
⅜						
½	0.632	0.314	0.097	0.125	0.166	0.188
¾	0.836	0.549	0.170	0.220	0.291	0.329
1	1.063	0.887	0.275	0.355	0.470	0.532
1¼	1.394	1.526	0.473	0.610	0.809	0.916
1½	1.624	2.071	0.642	0.829	1.098	1.243
2	2.083	3.408	1.056	1.363	1.806	2.045
2½	2.489	4.866	1.508	1.946	2.579	2.919
3	3.090	7.499	2.325	3.000	3.974	4.499
3½	3.570	10.010	3.103	4.004	5.305	6.006
4	4.050	12.882	3.994	5.153	6.828	7.729
5	5.073	20.212	6.266	8.085	10.713	12.127
6	6.093	29.158	9.039	11.663	15.454	17.495

(continued on next page)

DIMENSIONS AND PERCENT AREA OF CONDUIT AND TUBING

(For the combinations of wires permitted in Chapter 9, Table 1, *NEC*®)
(See *Ugly's* pages 173-174 for metric conversions.)

Trade Size Inches	Internal Diameter Inches	Total Area 100% Sq. Inches	2 Wires 31% Sq. Inches	Over 2 Wires 40% Sq. Inches	1 Wire 53% Sq. Inches	(NIPPLE) 60% Sq. Inches
Rigid PVC Conduit (PVC), Schedule 80						
½	0.526	0.217	0.067	0.087	0.115	0.130
¾	0.722	0.409	0.127	0.164	0.217	0.246
1	0.936	0.688	0.213	0.275	0.365	0.413
1¼	1.255	1.237	0.383	0.495	0.656	0.742
1½	1.476	1.711	0.530	0.684	0.907	1.027
2	1.913	2.874	0.891	1.150	1.523	1.725
2½	2.290	4.119	1.277	1.647	2.183	2.471
3	2.864	6.442	1.997	2.577	3.414	3.865
3½	3.326	8.688	2.693	3.475	4.605	5.213
4	3.786	11.258	3.490	4.503	5.967	6.755
5	4.768	17.855	5.535	7.142	9.463	10.713
6	5.709	25.598	7.935	10.239	13.56	15.359
Rigid PVC Conduit (PVC), Schedule 40 & HDPE Conduit (HDPE)						
½	0.602	0.285	0.088	0.114	0.151	0.171
¾	0.804	0.508	0.157	0.203	0.269	0.305
1	1.029	0.832	0.258	0.333	0.441	0.499
1¼	1.360	1.453	0.450	0.581	0.770	0.872
1½	1.590	1.986	0.616	0.794	1.052	1.191
2	2.047	3.291	1.020	1.316	1.744	1.975
2½	2.445	4.695	1.455	1.878	2.488	2.817
3	3.042	7.268	2.253	2.907	3.852	4.361
3½	3.521	9.737	3.018	3.895	5.161	5.842
4	3.998	2.554	3.892	5.022	6.654	7.532
5	5.016	19.761	6.126	7.904	10.473	11.856
6	6.031	28.567	8.856	11.427	15.141	17.140
Type A, Rigid PVC Conduit (PVC)						
½	0.700	0.385	0.119	0.154	0.204	0.231
¾	0.910	0.650	0.202	0.260	0.345	0.390
1	1.175	1.084	0.336	0.434	0.575	0.651
1¼	1.500	1.767	0.548	0.707	0.937	1.060
1½	1.720	2.324	0.720	0.929	1.231	1.394
2	2.155	3.647	1.131	1.459	1.933	2.188
2½	2.635	5.453	1.690	2.181	2.890	3.272
3	3.230	8.194	2.540	3.278	4.343	4.916
3½	3.690	10.694	3.315	4.278	5.668	6.416
4	4.180	13.723	4.254	5.489	7.273	8.234
Type EB, PVC Conduit (PVC)						
2	2.221	3.874	1.201	1.550	2.053	2.325
2½	–	–	–	–	–	–
3	3.330	8.709	2.700	3.484	4.616	5.226
3½	3.804	11.365	3.523	4.546	6.023	6.819
4	4.289	14.448	4.479	5.779	7.657	8.669
5	5.316	22.195	6.881	8.878	11.763	13.317
6	6.336	31.530	9.774	12.612	16.711	18.918

Source: NFPA 70®, National Electrical Code®, NFPA, Quincy, MA, 2020, Table 4, as modified.

METAL BOXES

Box Dimension, Inches Trade Size, or Type	Min. Cu. In. Capacity	Maximum Number of Conductors*						
		No. 18	No. 16	No. 14	No. 12	No. 10	No. 8	No. 6
4 x 1-1/4 Round or Octagonal	12.5	8	7	6	5	5	4	2
4 x 1-1/2 Round or Octagonal	15.5	10	8	7	6	6	5	3
4 x 2-1/8 Round or Octagonal	21.5	14	12	10	9	8	7	4
4 x 1-1/4 Square	18.0	12	10	9	8	7	6	3
4 x 1-1/2 Square	21.0	14	12	10	9	8	7	4
4 x 2-1/8 Square	30.3	20	17	16	13	12	10	6
4-11/16 x 1-1/4 Square	25.5	17	14	12	11	10	8	5
4-11/16 x 1-1/2 Square	29.5	19	16	14	13	11	9	5
4-11/16 x 2-1/8 Square	42.0	28	24	21	18	16	14	8
3 x 2 x 1-1/2 Device	7.5	5	4	3	3	3	2	1
3 x 2 x 2 Device	10.0	6	5	5	4	4	3	2
3 x 2 x 2-1/4 Device	10.5	7	6	5	4	4	3	2
3 x 2 x 2-1/2 Device	12.5	8	7	6	5	5	4	2
3 x 2 x 2-3/4 Device	14.0	9	8	7	6	5	4	3
3 x 2 x 3-1/2 Device	18.0	12	10	9	8	7	5	3
4 x 2-1/8 x 1-1/2 Device	10.3	6	5	5	4	4	3	2
4 x 2-1/8 x 1-7/8 Device	13.0	8	7	6	5	5	4	2
4 x 2-1/8 x 2-1/8 Device	14.5	9	8	7	6	5	4	2
3-3/4 x 2 x 2-1/2 Masonry Box/Gana	14.0	9	8	7	6	5	4	2
3-3/4 x 2 x 3-1/2 Masonry Box/Gana	21.0	14	12	10	9	8	7	4
FS-Minimum Internal Depth 1-3/4 Single cover/Gang	13.5	9	7	6	6	5	4	2
FD-Minimum Internal Depth 2-3/8 Single Cover/Grang	18.0	12	10	9	8	7	6	3
FS-Minimum Internal Depth 1-3/4 Multiple Cover/Gang	18.0	12	10	9	8	7	6	3
FD-Minimum Internal Depth 2-3/8 Multiple Cover/Gang	24.0	16	13	12	10	9	8	4

*Where no volume allowances are required by 314.16(B)(2) through (B)(5).
Source: NFPA *70*, *National Electrical Code* ®. NFPA, Quincy, MA, 2020, Table 314.16(A), as modified.

VOLUME REQUIRED PER CONDUCTOR

Size of Conductor	Free Space Within Box for Each Conductor
No. 18	1.5 Cubic Inches
No. 16	1.75 Cubic Inches
No. 14	2 Cubic Inches
No. 12	2.25 Cubic Inches
No. 10	2.5 Cubic Inches
No. 8	3 Cubic Inches
No. 6	5 Cubic Inches

For complete details, see *NEC* 314.16(B).
Source: NFPA 70®, *National Electrical Code*®, NFPA, Quincy, MA, 2020, Table 314.16(B), as modified.

BOX FILL—-SUMMARY OF CONTRIBUTING ITEMS

Items Within Box	Allowance	Based On*
Each conductor originating outside box and terminated or spliced within the box	One for each conductor	Actual conductor size
Each conductor passing through box without splice or termination (less than 12 in. total length)	One for each unbroken conductor	Actual conductor size
Conductors—coiled (or looped) and unbroken that are 12 in. or greater	Two for each single, unbroken conductor	Actual conductor size
Conductors—originating within box and not leaving box	None (these not counted)	N/A
One or more internal cable clamp(s)	One only	Largest-sized conductor present
Support fittings	One for each kind of support fitting	Largest-sized conductor present
Devices or utilization equipment	Two for each yoke or mounting strap and two for each gang needed for mounting devices wider than 2 in.	Largest-sized conductor connected to device or utilization equipment

(continued on next page)

BOX FILL—-SUMMARY OF CONTRIBUTING ITEMS

Items Within Box	Allowance	Based On*
Up to 4 equipment grounding conductors	One only	Largest equipment grounding conductor
Each additional equipment grounding conductor	1/8 volume allowance	Largest equipment grounding conductor

Note: An equipment grounding conductor or conductors of not over 4 fixture wires smaller than 14 AWG, or both, can be omitted from the calculations where they enter a box from a domed luminaire or similar canopy and terminate within that box. [314.16(B)(1) Exception]
* Refer to *NEC*® Table 314.16(B).

 BOX FILL CALCULATION

This Device Box Contains Components and Conductors Requiring Deductions in Accordance with *NEC*® 314.16

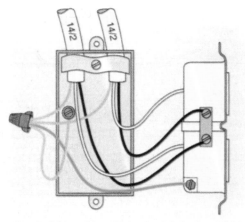

Standard 3 in. × 2 in. × 3½ in. metal device box (18.0 in.³)

BOX FILL CALCULATION

Box with duplex receptacle and conductors all the same size. Determine whether a 3 x 2 x 3½ in. (75 x 50 x 90 mm) device box (18 in.^{3} or 295 cm^3) is large enough to contain a duplex receptacle and four 14 AWG conductors.

- The following table shows the volume allowances for the various items installed in this box.
- The total box volume required is 16 in.^{3} (262.4 cm^3).
- The volume of the metal box installed is 18 in.^{3} (295 cm^3).
- This box complies with the box fill calculation requirements in 314.16.

Total Box Fill for Box with Duplex Receptacle and Conductors All the Same Size

Items Contained in Box	Volume Allowance	Unit Volume* in.^{3} (cm^3)	Total Box Fill in.^{3} (cm^3)
Four conductors	Four volume allowances for 14 AWG conductors	2.00 (32.8)	8.00 (131.2)
One cable clamp	One volume allowance (based on 14 AWG conductors)	2.00 (32.8)	2.00 (32.8)
One device	Two volume allowances (based on 14 AWG conductors)	2.00 (32.8)	4.00 (65.6)
Equipment grounding conductors (all)	One volume allowance (based on 14 AWG conductors)	2.00 (32.8)	2.00 (32.8)
Total			16.00 (262.4)

* Based on *NEC*® Table 314.16(B).

🔌 JUNCTION BOX SIZING

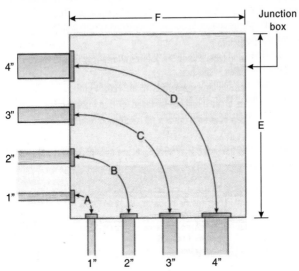

A—6 times conduit size = 6" minimum
B—6 times conduit size = 12" minimum
C—6 times conduit size = 18" minimum
D—6 times conduit size = 24" minimum
E—6 times largest conduit size, plus
    all other conduits entering
    6 x 4 + 3 + 2 + 1 = 30" minimum
F—in this case, same as E

© Jones & Bartlett Learning.

CLEAR WORKING SPACE IN FRONT OF ELECTRICAL EQUIPMENT (MINIMUM DEPTH)

Nominal Voltage to Ground	Conditions		
	1	2	3
	Minimum Clear Distance (ft)		
0–150V	3	3	3
151-600V	3	3½	4
601-2500 V	3	4	5
2501-9000 V	4	5	6
9001-25000 V	5	6	9
25001 V-75 kV	6	8	10
Above 75 kV	8	10	12

Notes:
1. For SI units, 1 foot = 0.3048 meter.
2. Where the conditions are as follows:
   Condition 1—Exposed live parts on one side of the working space and no live or grounded parts on the other side of the working space, or exposed live parts on both sides of the working space that are effectively guarded by insulating materials.
   Condition 2—Exposed live parts on one side of the working space and grounded parts on the other side of the working space. Concrete, brick, or tile walls shall be considered as grounded.
   Condition 3—Exposed live parts on both sides of the working space.

See *Ugly's* pages 173-174 for metric conversions. For electrical rooms where the equipment is rated 800 amps or more, A/EC110.26(C)(3) requires personnel doors to open in the egress direction and be equipped with listed panic hardware or listed fire exit hardware. Entrance to rooms shall meet the requirements of 110.27(C) and shall be clearly marked with warning signs forbidding unqualified persons to enter.

Source: NFPA 70®, *National Electrical Code®*, NFPA, Quincy, MA, 2020, Tables 110.26(A)(1) and 110.34(A), as modified.

MINIMUM COVER REQUIREMENTS, 0 TO 1000 VOLTS, NOMINAL

For 0 to 1000 Volts, Nominal, Burial in Inches

Location of Circuit or Wiring Method	Type of Circuit or Wiring Method				
	Column 1 Direct Burial Conductors or Cables	Column 2 Intermediate Metal Conduit or Rigid Metal Conduit	Column 3 Nonmetallic Raceways Listed for Direct Burial (Without Concrete Encasement or Other Approved Raceways)	Column 4 Residential Branch Circuits Rated 120 Volts or Less with GFCI and Maximum Overcurrent Protection of 20 Amperes	Column 5 Circuits for Control of Landscape Lighting and Limited to 30 Volts Maximum and Installed with Type UF (or in Other Identified Cable or Raceway)
	Inches	Inches	Inches	Inches	Inches
All locations except where specified below	24	6	18	12	6
In trench below 2-in. thick concrete or similar material	18	6	12	6	6
Under minimum 4-in. thick concrete exterior slab with no car traffic and the slab extending at least 6 in. beyond underground installation	18	4	4	6 (direct burial 4 (in raceway only)	6 (direct burial 4 (in raceway only)

(continued on next page)

Under roads, highways, streets, driveways, parking lots, and alleys	24	24	24	24	24
Driveways and outdoor parking areas for one- and two-family dwellings, used only for dwelling-related purposes	18	18	18	12	18

Notes:
1. Cover is defined as shortest distance in inches measured between a spot on the top surface of any direct-buried conductor, cable, conduit (or other raceway) and the top surface (e.g., finished grade, concrete, or similar cover).
2. Raceways approved for burial only where concrete encased require concrete envelope at least 2 in. thick.
3. Smaller depths shall be permitted where conductors and cables rise for splices or terminations or where access is otherwise required.
4. Where one of the wiring methods listed in Columns 1–3 is used for one of the circuit types in Columns 4 and 5, the shallowest burial depth is permitted.
5. If the cover depths listed here are prevented by solid rock, the wiring shall be installed in metal or nonmetallic raceway permitted for direct burial. These raceways shall be covered by at least 2 in. of concrete extending down to the rock.

Adapted from *NEC*® Table 300.5.

RESIDENTIAL GROUNDING DIAGRAMS

This Grounding Electrode System Uses the Metal Frame of a Building, a Ground Ring, a Concrete-Encased Electrode, a Metal Underground Water Pipe, and a Ground Rod

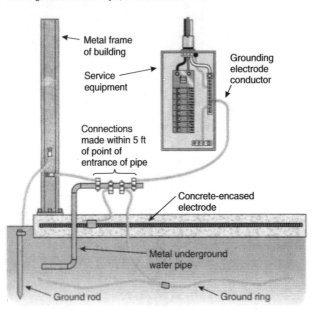

Metal frame of building

Service equipment

Grounding electrode conductor

Connections made within 5 ft of point of entrance of pipe

Concrete-encased electrode

Metal underground water pipe

Ground rod

Ground ring

© Jones & Bartlett Learning.

🔌 RESIDENTIAL GROUNDING DIAGRAMS

This Service Equipment Is Grounded to Metal Water Piping Within 5 ft (1.52 m) of Entry Into a Dwelling

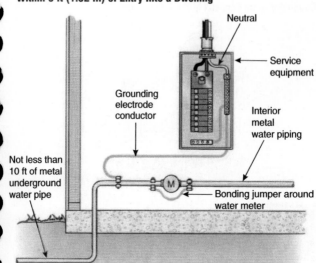

Note: a metal underground water pipe shall be supplemented by an additional electrode of a type specified in 250.52(A)(2) through (A)(8). See 250.53(D)(2).

© Jones & Bartlett Learning.

RESIDENTIAL GROUNDING DIAGRAMS

Use a Listed Clamp for Connecting a Grounding Electrode
Conductor to an Exposed Metal Water Pipe

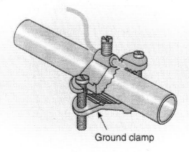

Ground clamp

RESIDENTIAL GROUNDING DIAGRAMS

Bonding Jumpers at a Gas-Fired Water Heater Must Be Long Enough to Permit Removal of the Equipment Without Losing the Integrity of the Grounding Path

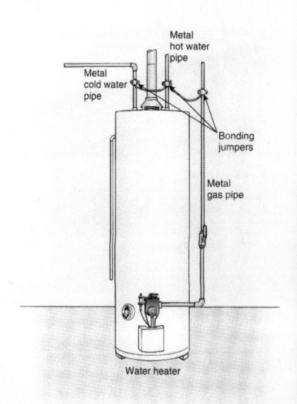

Metal hot water pipe

Metal cold water pipe

Bonding jumpers

Metal gas pipe

Water heater

33

RESIDENTIAL GROUNDING DIAGRAMS

The Installation Requirements for Rod and Pipe Electrodes Are Specified in *NEC* 250.53(A)(4)

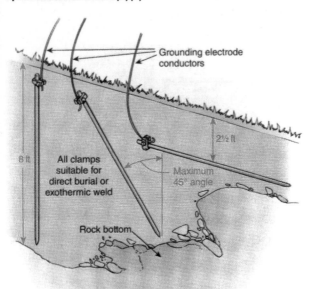

Grounding electrode conductors

2½ ft

8 ft

All clamps suitable for direct burial or exothermic weld

Maximum 45° angle

Rock bottom

© Jones & Bartlett Learning.

34

Minimum 6 ft (1.8 m) Spacing Between Grounding Electrodes Is Required by *NEC* 250.53(A)(3)

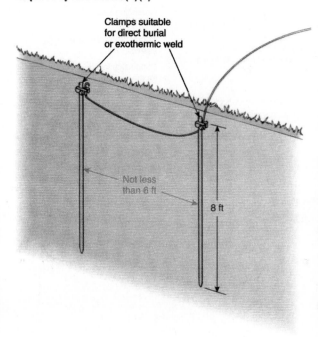

Clamps suitable for direct burial or exothermic weld

Not less than 6 ft

8 ft

© Jones & Bartlett Learning.

Concrete-encased Electrodes Are Often Steel Reinforcing Bars (Rebar)

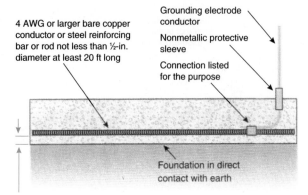

4 AWG or larger bare copper conductor or steel reinforcing bar or rod not less than ½-in. diameter at least 20 ft long

Grounding electrode conductor

Nonmetallic protective sleeve

Connection listed for the purpose

Foundation in direct contact with earth

2 in. min.

© Jones & Bartlett Learning.

🔌 RESIDENTIAL GROUNDING DIAGRAMS

A Metal Raceway that Contains a Grounding Electrode Conductor Is Required to Be Bonded to the Conductor at Both Ends by *NEC* 250.64(E)

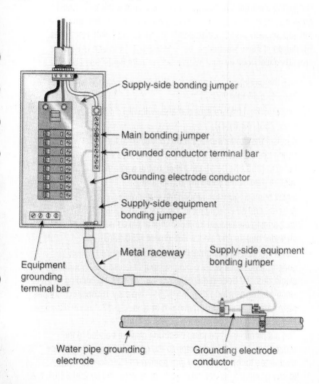

Supply-side bonding jumper

Main bonding jumper

Grounded conductor terminal bar

Grounding electrode conductor

Supply-side equipment bonding jumper

Metal raceway

Supply-side equipment bonding jumper

Equipment grounding terminal bar

Water pipe grounding electrode

Grounding electrode conductor

© Jones & Bartlett Learning.

🔌 GROUNDING ELECTRODE SELECTION

Section 250.50 requires all grounding electrodes as described in 250.52(A)(1) through (A)(7) that are present at each building or structure served to be bonded together to form a grounding electrode system. Grounding electrodes include metal underground water pipe, metal in-ground support structure(s), concrete-encased electrodes, ground ring, rod and pipe electrodes, other listed electrodes, and plate electrodes. For details of these electrodes, see 250.52(A)(1) through (A)(7). For grounding electrode system installations requirements, see 250.53.

- Size the grounding electrode conductor in accordance with *NEC*® 250.66 and Table 250.66.
- Size bonding jumpers in accordance with *NEC*® 250.102(C)(1) and Table 250[102(C)(1)],
- Connect the grounding electrode conductor to the metal water piping system within 5 ft (1.52 m) of the place where it enters the building [250.68(C)(1)] and ahead of the water meter [250.68(B)], using a pipe clamp listed for the purpose.
- Install bonding jumpers around the water meter, water heater, filtration devices, and other equipment. Continuity of the grounding path and the bonding connection to interior water piping is not permitted to rely on water meters and similar equipment [250.53(D)(1)].
- Install a supplemental grounding electrode if needed. When underground metal water piping systems are used as grounding electrodes, the Code requires that a supplemental grounding electrode be installed [250.53(D)(2)]—typically, a copper or copper-plated steel ground rod a minimum of ⅝ in. (15.87 mm) in diameter and at least 8 ft (2.44 m) long [250.52(A)(5)],
  The supplemental grounding electrode may be bonded to the grounding electrode conductor, to the grounded service-entrance conductor, to the nonflexible grounded metal service raceway, to any grounded service enclosure, or as provided by 250.32(B). It is not required to be larger than 6 AWG copper or 4 AWG aluminum [250.53(E)],

 GROUNDING ELECTRODE SELECTION

- Install multiple supplemental grounding electrodes if necessary. Where more than one rod or pipe electrode is used, they shall not be less than 6 ft (1.83 m) from any other electrode [250.53(B)], Only one supplemental rod or pipe electrode is required if the electrode has a resistance to earth of 25 ohms or less.

GROUNDING ELECTRODE CONDUCTOR FOR ALTERNATING-CURRENT SYSTEMS

Size of Largest Ungrounded Conductor or Equivalent Area for Parallel Conductors (AWG/kcmil)		Size of Grounding Electrode Conductor (AWG/kcmil)	
Copper	Aluminum or Copper-Clad Aluminum	Copper	Aluminum or Copper-Clad Aluminum
2 or Smaller	1/0 or Smaller	8	6
1 or 1/0	2/0 or 3/0	6	4
2/0 or 3/0	4/0 or 250	4	2
Over 3/0 through 350	Over 250 through 500	2	1/0
Over 350 through 600	Over 500 through 900	1/0	3/0
Over 600 through 1100	Over 900 through 1750	2/0	4/0
Over 1100	Over 1750	3/0	250

Notes:
1. If multiple sets of service-entrance conductors connect directly to a service drop, set of overhead service conductors, set of underground service conductors, or service lateral, the equivalent size of the largest service-entrance conductor shall be determined by the largest sum of the areas of the corresponding conductors of each set.
2. Where there are no service-entrance conductors, the grounding electrode conductor size shall be determined by the equivalent size of the largest service-entrance conductor required for the load to be served.
3. See installation restrictions in *NEC* 250.64(A).
Source: NFPA 70®, National Electrical Code®, NFPA, Quincy, MA, 2020, Table 250.66, as modified.

GROUNDED CONDUCTOR, MAIN BONDING JUMPER, SYSTEM BONDING JUMPER, AND SUPPLY-SIDE BONDING JUMPER FOR ALTERNATING-CURRENT SYSTEMS

Size of Largest Ungrounded Conductor or Equivalent Area for Parallel Conductors (AWG/kcmil)		Size of Grounded Conductor or Bonding Jumper* (AWG/kcmil)	
Copper	Aluminum or Copper-Clad Aluminum	Copper	Aluminum or Copper-Clad Aluminum
2 or Smaller	1/0 or Smaller	8	6
1 or 1/0	2/0 or 3/0	6	4
2/0 or 3/0	4/0 or 250	4	2
Over 3/0 through 350	Over 250 through 500	2	1/0
Over 350 through 600	Over 500 through 900	1/0	3/0
Over 600 through 1100	Over 900 through 1750	2/0	4/0
Over 1100	Over 1750	See Notes 1 and 2	

Notes:

1. If the ungrounded supply conductors are larger than 1100 kcmil copper or 1750 kcmil aluminum, the grounded conductor or bonding jumper shall have an area not less than 121½ percent of the area of the largest ungrounded supply conductor or equivalent area for parallel supply conductors. The grounded conductor or bonding jumper shall not be required to be larger than the largest ungrounded conductor or set of ungrounded conductors.
2. If the ungrounded supply conductors are larger than 1100 kcmil copper or 1750 kcmil aluminum and if the ungrounded supply conductors and the bonding jumper are of different materials (copper, aluminum, or copper-clad aluminum), the minimum size of the grounded conductor or bonding jumper shall be based on the assumed use of ungrounded supply conductors of the same material as the grounded conductor or bonding jumper and will have an ampacity equivalent to that of the installed ungrounded supply conductors.
3. If multiple sets of service-entrance conductors are used as permitted in 230.40, Exception No. 2, or if multiple sets of ungrounded supply conductors are installed for a separately derived system, the equivalent size of the largest ungrounded supply conductor(s) shall be determined by the largest sum of the areas of the corresponding conductors of each set.
4. If there are no service-entrance conductors, the supply conductor size shall be determined by the equivalent size of the largest serviceentrance conductor required for the load to be served.
*For the purposes of applying this table and its notes, the term *bonding jumper* refers to main bonding jumpers, system bonding jumpers, and supply-side bonding jumpers.
Source: NFPA 70®, *National Electrical Code*®, NFPA, Quincy, MA, 2020, Table 250.102(C)(1), as modified

MINIMUM SIZE EQUIPMENT GROUNDING CONDUCTORS FOR GROUNDING RACEWAY AND EQUIPMENT

Rating or Setting of Automatic Overcurrent Device in Circuit Ahead of Equipment, Conduit, Etc., Not Exceeding (Amperes)	Size (AWG or kcmil)	
	Copper	Aluminum or Copper-Clad Aluminum*
15	14	12
20	12	10
60	10	8
100	8	6
200	6	4
300	4	2
400	3	1
500	2	1/0
600	1	2/0
800	1/0	3/0
1000	2/0	4/0
1200	3/0	250 kcmil
1600	4/0	350 kcmil
2000	250 kcmil	400 kcmil
2500	350 kcmil	600 kcmil
3000	400 kcmil	600 kcmil
4000	500 kcmil	750 kcmil
5000	700 kcmil	1200 kcmil
6000	800 kcmil	1200 kcmil

Note: Where necessary to comply with NEC250.4(A)(5) or 250.4(B)(4), the equipment grounding conductor shall be sized larger than given in this table.
* See installation restrictions in *NEC* 250.120.
Source: NFPA 70®, *National Electrical Code*®, NFPA, Quincy, MA, 2020, Table 250.122, as modified.

Connection of Lamp Bases

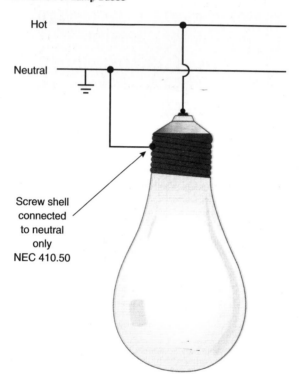

Hot

Neutral

Screw shell
connected
to neutral
only
NEC 410.50

© Jones & Bartlett Learning.

 RACEWAY INSTALLATION GUIDELINES

- Raceways should be cut carefully and the ends reamed to create a smooth surface that will not damage conductor insulation when wires are pulled into them. Where raceways contain 4 AWG or larger insulated circuit conductors, and these conductors enter a cabinet, a box, an enclosure, or a raceway, the conductors shall be protected by an identified fitting providing a smoothly rounded insulating surface or one of the other options in 300.4(G)(2), (3), or (4).

- Raceways must be securely supported, at intervals that vary from one type to another. Horizontal runs of raceways can be supported by openings in framing members.

- Bends must be carefully made so that raceways are not damaged and the internal cross-sectional area is not reduced. There can be a maximum of four right-angle bends (a total of 360 degrees) between pull points such as boxes, conduit bodies, and panelboard cabinets. This total includes offsets at boxes and enclosures. An offset with two 10-degree bends counts as 20 degrees [348.26 (flexible metal conduit: Type FMC), 350.26 (liquidtight flexible metal conduit: Type LFMC), 352.26 (rigid polyvinyl chloride conduit: Type PVC), 356.26 (liquidtight flexible nonmetallic conduit: Type LFNC), 358.26 (electrical metallic tubing: Type EMT), and 362.26 (electrical nonmetallic tubing: Type ENT)].

- Generally, raceway systems must be installed completely between outlets and other pull points before the conductors are pulled into them [300.18(A)].

SUPPORTS FOR RIGID METAL CONDUIT

Conduit Size	Distance Between Supports
½ in.–¾ in.	10 feet
1 in.	12 feet
1¼ in-1½ in.	14 feet
2 in.-2½ in.	16 feet
3 in. and larger	20 feet

Source: NFPA 70®, *National Electrical Code*®, NFPA, Quincy, MA, 2020, Table 344.30(B)(2), as modified.

SUPPORTS FOR RIGID NONMETALLIC CONDUIT

Conduit Size	Distance Between Supports
½ in.–1 in.	3 feet
1¼ in. –2 in.	5 feet
2 ½ in.-3 in.	6 feet
3½ in. -5 in.	7 feet
6 in.	8 feet

For SI units: (Supports) 1 foot = 0.3048 meter.
Source: NFPA70®, *National Electrical Code*®, NFPA, Quincy, MA, 2020, Table 352.30, as modified.

ELECTRICAL METALLIC TUBING—EMT

Type EMT is covered in Article 358 in the *NEC*®. EMT is an unthreaded thinwall raceway manufactured in both steel and aluminum, though the steel version is more common. It is joined with a threadless set screw or compression fittings.

Uses. EMT can be used in wet locations both indoors and outdoors, either exposed or concealed, and can be cast in concrete. It cannot be used to support luminaires or other equipment. Electrical equipment (of any type) is considered concealed when it is made inaccessible by the structure or finish of the building.

⚡ ELECTRICAL METALLIC TUBING—EMT

Installation. EMT can be cut with a hacksaw and the end reamed after cutting. Bends should be made using a conduit bender to avoid deforming the tubing. EMT must be securely fastened in place every 10 ft (3.0 m) and within 3 ft (900 mm) of a box, cabinet, or other termination, but is permitted to be unsupported in accordance with one of the two exceptions [358.30(A)]. Horizontal runs of EMT supported by openings through framing members at intervals not greater than 10 ft (3 m) and securely fastened within 3 ft (900 mm) of termination points shall be permitted [358.30(B)].

Raceway Sizes and Numbers of Conductors. Table C.1 of *NEC*® Annex C is used to determine the maximum conductor fill for EMT when all wires are the same size. (EMT is one of the raceways commonly used in residential wiring.) Use Tables 4 and 5 of NEC Chapter 9 for mixed conductor sizes.

Grounding. Electrical metallic tubing is permitted to serve as an equipment grounding conductor for circuits installed within it, and when it forms a complete wiring system with metal boxes [250.118(4)]. However, in many applications, a separate equipment grounding conductor is run for convenience.

Offset Bends—Using a Hand-Bender

An offset bend is used to change the level or plane of the conduit. This is usually necessitated by the presence of an obstruction in the original conduit path.

Step One:

Determine the offset depth. (X)

⚡ ELECTRICAL METALLIC TUBING—EMT

Step Two:

Multiply the offset depth "x" the multiplier for the degree of bend used to determine the distance between bends.

ANGLE		MULTIPLIER
10° x 10°	=	6
22½° x 22½°	=	2.6
30° x 30°	=	2
45° X 45°	=	1.4
60° X 60°	=	1.2

Example: If the offset depth required (X) is 6 in. and you intend to use 30-degree bends, then the distance between bends is 6 in. x 2 = 12 in.

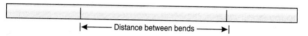

|←——— Distance between bends ———→|

Step Three:

Mark at the appropriate points, align the arrow on the bender with the first mark, and bend to desired degree by aligning EMT with chosen degree line on bender.

Step Four:

Slide down the EMT, align the arrow with the second mark, and bend to the same degree line. Be sure to note the orientation of the bender head. Check alignment.

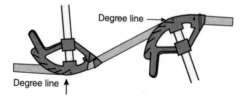

Degree line

Degree line ↑

46

ELECTRICAL METALLIC TUBING—EMT

90-degree Bends—Using a Hand-Bender

The stub is the most common bend.

Step One:

Determine the height of the stub-up required, and mark on EMT.

Step Two:

Find the "Deduct" or "Take-up" amount from the Take-Up Chart.

Subtract the take-up amount from the stub height, and mark the EMT that distance from the end.

Step Three:

Align the arrow on the bender with the last mark made on the EMT, and bend to the 90-degree mark on the bender

Description		Take-Up
½ in. EMT	=	5 in.
¾ in. EMT	=	6 in.
1 in. EMT	=	8 in.
1 ¼ in. EMT	=	11 in.

Deduct X
Height of stub

Height of stub

🔧 ELECTRICAL METALLIC TUBING—EMT

Back-to-Back Bends—Using a Hand-Bender

A back-to-back bend results in a "U" shape in a length of conduit. It is used for a conduit that runs along the floor or ceiling and turns up or down a wall.

Step One:

After the first 90-degree bend is made, determine the back-to-back length and mark on EMT.

Step Two:

Align this back-to-back mark with the star mark on the bender, and bend to 90 degrees.

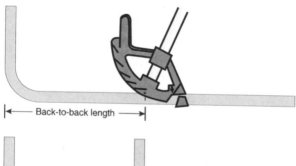

|←——— Back-to-back length ———→|

Completed bend

Three-Point Saddle Bends—Using Hand-Bender

The 3-point saddle bend is used when encountering an obstacle (usually another pipe).

Step One:

Measure the height of the obstruction.
Mark the center point on EMT.

Step Two:

Multiply the height of the obstruction by 2.5 and mark this distance on each side of the center mark.

Step Three:

Place the center mark on the saddle mark or notch. Bend to 45 degrees.

Step Four:

Bend the second mark to a 22½-degree angle at arrow.

Step Five:

Bend the third mark to a 22½-degree angle at arrow. Be aware of the orientation of the EMT on all bends. Check alignment

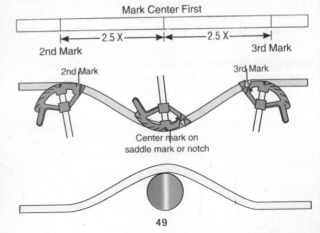

Mark Center First

←——2.5 X——→|←——2.5 X——→|

2nd Mark
3rd Mark

2nd Mark
3rd Mark

Center mark on
saddle mark or notch

49

FLEXIBLE METAL CONDUIT (FMC) RULES

FMC [348] is used where flexibility is needed, either for ease of installation or to provide vibration isolation at connections to motorized equipment. Wires are pulled in later.

Uses. Only indoors in dry locations, exposed or concealed. Trade size ½ is the smallest for general use. Trade size 3/8 can be used, in lengths up to 6 ft (1.8 m), to connect single appliances or luminaires [348.20(A)(2)],

Installation. Can be cut with a hacksaw. Attach to boxes, cabinets, and other equipment using special connectors. Must be securely fastened in place every 4½ ft (1.4 m) and within 12 in. (300 mm) of a box, cabinet, or other termination. Horizontal runs of FMC can be supported by holes or notches in framing members [348.30(B)], FMC can be fished without securing or supporting it. Unsupported lengths up to 3 ft (900 mm) are permitted where flexibility is needed. Unsupported lengths up to 6 ft (1.8 m) are permitted as whips to luminaires [348.30(A), Exceptions No. 1, 2, 3, and 4],

Protection from Physical Damage. FMC run parallel to studs and other framing wood members must comply with rules for cable wiring methods. Cable's edge must be $\geq$1¼ in. (32 mm) to any area where screws or nails might penetrate [300.4(D)], Otherwise, the conduit must be protected by a steel plate, sleeve, or other guard at least $^1/_{16}$ in. (1.6 mm) thick.

Raceway Sizes and Numbers of Conductors. Use Table C.3 of *NEC* Annex C to determine the maximum conductor fill for FMC when all wires are the same size. Use Tables 4 and 5 of NEC Chapter 9 for mixed conductor sizes.

Grounding. FMC cable may be used as grounding conductor in lengths up to 6 ft (1.8 m) with conductors protected at $\leq$ 20 A. Install separate ground wire when required for all other applications [250.118(5) and 348.60],

🔌 LIQUIDTIGHT FLEXIBLE METAL CONDUIT (LFNC) RULES

LFMC [350] is similar to flexible metal conduit with a liquidtight, sunlight-resistant outer jacket.

Uses. For outdoor connections to equipment. Trade size ½ is smallest for general use. Trade size 3/8 in. can be used, in lengths up to 6 ft (1.8 m), for single connections [350.20(A), Exception],

Oirect-bury where listed and marked for the purpose—needs $\geq$ 24 in. (610 mm) of cover unless it satisfies *NEC®* Table 300.5. LFMC may enclose accessory structure's feeder conductors.

Installation. Cut with a hacksaw. May bend by hand—minimum radii must comply with Other Bends column of *NEC®* Table 2, Chapter 9. Attach to boxes, cabinets, and other equipment with special connectors. Fasten every 4½ ft (1.4 m), within 12 in. (300 mm) of a box, cabinet, or other termination [350.30(A)]. Horizontal runs can be supported by holes/notches in framing members [350.30(B)],

Can be fished without securing or supporting. Unsupported lengths $\leq$ 3 ft (900 mm; for trade sizes ½-1¼), 4 ft (1.2 m; for trade sizes 1½-2), and 5 ft (1.5 m; for trade sizes 2¼ and larger) are permitted where flexibility is needed. Unsupported lengths up to 6 ft (1.8 m) are permitted as whips to luminaires [350.30(A), Exceptions No. 1,2, 3, and 4],

Protection from Physical Damage. LFMC run parallel to studs and other framing wood members must comply with rules for cable wiring methods. Edge of cable must be no closer than 1¼ in. (32 mm) to anywhere screws or nails are likely to penetrate [300.4(D)], Otherwise, the conduit must be protected by a steel plate, sleeve, or other guard at least $1/16$ in. (1.6 mm) thick.

Raceway Sizes and Numbers of Conductors. Table C.8 of *NEC®* Annex C gives maximum conductor fill for same-size wires. For mixed conductor sizes, use Tables 4 and 5 of *NEC®* Chapter 9.

Grounding. LFMC may serve as an equipment grounding conductor in lengths up to 6 ft (1.8 m) with conductors protected at 20 A or less. A separate ground wire must be installed when required for all other applications [250.118(6) and 350.60].

LIQUIDTIGHT FLEXIBLE NONMETAL CONDUIT (LFNC) RULES

LFNC [356] is similar to LFMC, except completely nonmetallic. There are three types:

- Type LFNC-A has strength reinforcement between the core and cover.
- Type LFNC-B has reinforcement built into the smooth conduit wall.
- Type LFNC-C is of corrugated construction, without integral reinforcement.

Uses. LFNC-B (FNMC is an alternate designation for LFNC) [356.2(3), Informational Note]) used for outdoor connections to motorized equipment. Trade size ½ is smallest used for most applications. Direct-bury (needs ≥ 24 in. (610 mm) of cover, unless it satisfies 300.5) or encase in concrete where listed and marked for the purpose. LFNC may enclose accessory building's feeder conductors.

Installation. Cut with a hacksaw. May bend by hand—minimum radii must satisfy "Other Bends" column of *NEC®* Chapter 9 Table 2. Fasten every 3 ft (900 mm), ≤ 12 in. (300 mm) of box, cabinet, or other termination. Holes/notches in framing members may support horizontal runs [356.30(3)].

Can be fished without securing or supporting. Unsupported lengths ≤ 3 ft (0.9 m) allowed where flexibility is needed; ≤ 6 ft (1.8 m) as whips to luminaires [356.30(1), (2), and (4)].

Protection from Damage. Runs parallel to studs and framing wood members must follow cable wiring methods rules. Cable's edge must be no closer than 1¼ in. (32 mm) from areas that screws or nails might pierce [300.4(D)] or protect conduit with steel plate, sleeve, or guard ≥ 1/16 in. (1.6 mm) thick.

Raceway Sizes and Numbers of Conductors. Table C.5 of NEC Annex C gives maximum conductor fill for same-size wires. For mixed conductor sizes, use Tables 4 and 5 of *NEC®* Chapter 9.

LIQUIDTIGHT FLEXIBLE NONMETAL CONDUIT (LFNC) RULES

Grounding. Install equipment grounding conductor in LFNC when required for the application.

Prewired LFNC-B. Available as a prewired assembly with factory-installed conductors—more common for commercial applications.

NONMETALLIC SHEATHED CABLE: TYPE NM AND NMC—ROMEX

Article 334 of *NEC*

- Sizes are 14 AWG through 2 AWG copper conductors.
- Cable contains equipment ground—may be bare or green.
- Type NM can be installed exposed or concealed in dry locations.
- Type NMC can be installed exposed or concealed in dry, moist, or damp locations.
- May not be embedded in wet cement or used as service cable.
- Must be supported every 4½ ft (1.4 m) and within 12 in. (300 mm) of every box, fitting, or panel.
- Type NMC cables may be installed in corrosive locations, except as prohibited by 330.10(3).

NM CABLE RULES

Securing. Secure at intervals of 4½ ft (1.4 m), and 12 in. (300 mm) from every box, cabinet, or termination. Use staples, cable ties listed and identified for securement and support, or straps, hangers, or similar fittings designed and installed so as not to damage the cable. Two cables run together, one on top of the other, must be stapled flat when secured with the same staple or tie [334.30],

 NM CABLE RULES

Horizontal runs through holes/notches in framing members (wall studs, ceiling joists), spaced 4½ ft (1.4 m) apart, have adequate support. Ties unneeded for runs through framing members, but cable must also be secured within 12 in. (300 mm) of each box [334.30(A)],

Cable can be used unsupported where fished, installed in panels for prefabricated houses, and in whips of 4½ ft (1.4 m) for equipment installed within accessible ceilings [334.30(B)].

Plastic Device Boxes. Cable need not be secured to a single-gang nonmetallic box if it is fastened within 8 in. (200 mm) of the box. Multiple cables may enter one cable knockout opening [314.17(B)],

In Dropped or Suspended Ceilings. Cable installed above suspended ceilings must be installed as per 300.11 (B).

Bending Radius. Radius of the curved inner edge of a bend must be no less than 5 x cable diameter [334.24].

Protection Against Damage (Wood Framing). Install cable, parallel to framing members, with edges 1¼ in. (32 mm) from areas where screws or nails might penetrate—applies to vertical framing members (e.g., studs) [300.4(D)], Otherwise, protect cable with a steel plate, sleeve, or other guard $^1/_{16}$ in. (1.6 mm) thick, or one providing equal protection [300.4(A)(1) and Exception 2],

For multiple cables run through one hole in framing members near thermal insulation, caulk, or foam, reduce ampacities of conductors as per Table 310.15(C)(1). For 4-6 current-carrying conductors, reduce to 80% of the rated ampacity. For 7-9, reduce to 70% [334.80].

Drill holes through vertical wood studs or other framing members 1¼ in. (32 mm) from nearest edge, or provide a steel protector [300.4(A)].

 **NM CABLE RULES**

Protection Against Damage (Metal Framing). Cover all metal edges with listed bushings or grommets [300.4(B)(1)], Factory-made openings in steel studs keep cables 1¼ in. (32 mm) from front edge of stud. Field-cut holes follow rules for holes in wood framing members.

Protection Against Damage (Through Floors). Enclose cable in rigid metal conduit, intermediate metal conduit, EMC, Schedule 80 PVC rigid nonmetallic conduit, or other approved means extending no less than 6 in. (150 mm) above the floor [334.15(B)].

Keep cables, passing through floor plates in void spaces of walls, 1¼ in. (32 mm) from edges of framing members.

Boxes or conduit bodies are not needed at ends of raceways protecting cable. A fitting is needed on end(s) of conduit or tubing [300.15(C)]. Short metal raceways protecting cable need not be grounded [250.86, Exception 2], Consult AHJ over whether grounding is required.

In Exposed Locations. Cable must run along building finish surface or be on running boards.

- *Unfinished basements.* Sizes 6/2 or 8/3 and larger can be stapled directly to lower edges of joists. Run smaller cables through holes or on running boards [334.15(C)].
- *Accessible attics.* Protect cables run atop floor joists with guard strips, for attics with stairs. For a scuttle hole without stairs, install guard strip within 6 ft (1.8 m) of the hole's edge. Guard strips aren't needed for runs parallel to sides of rafters, joists, or studs [334.23].

Keep cables, run parallel to framing members or through bored holes, back 1¼ in. (32 mm) from face of stud or other framing member [300.4(A)] unless supplemental protection is provided.

RULES FOR NM CABLES AND BOXES

The *Code* rules for NM cables entering boxes can be summarized as follows:

- When nonmetallic-sheathed cables enter boxes, the sheath must extend into the box at least ¼ in. (6 mm) so that unprotected conductors are not exposed to damage [314.17(B)(1)],

- Cable connectors used with Type NM cables can be either separate fittings or connector devices that are an integral part of boxes. Multiple cables are permitted to enter a box through a single opening [314.17(B)(2), Exception],

- Nonmetallic-sheathed cables must be secured to boxes. Separate metal connectors are used when the cable enters metal boxes. Most plastic boxes are manufactured with integral cable clamps. These clamps are small, springy flaps through which the NM cable is pushed into the box. The clamps prevent the cable from being pulled out again.

- Type NM cable is permitted to enter a nonmetallic box no larger than 2¼ x 4 in. (57 x 100 mm) (a single-gang device box) without being secured when it is fastened within 8 in. (200 mm) of the box. In other words, NM cable can enter a box without a connector if the cable is fastened closer to the box than usual [314.17(B)(2), Exception],

- At least 6 in. (150 mm) of free conductor (measured from the end of the nonmetallic sheath) must be left at each box for splices of the connections of switches, receptacles, and luminaires. Each conductor must also be long enough to extend 3 in. (75 mm) outside the box. Although these are the *Code* minimums specified in 300.14, leaving longer wire ends at boxes and cabinets often makes termination much easier.

ARMORED CABLE: TYPE AC - BX

Article 320 of *NEC*®

- Contains bonding strip for bonding.
- Insulating bushing or its equivalent protection must be installed between armor and conductors. The connector or clamp by which the Type AC cable is fastened to boxes or cabinets shall be of such design that the insulating bushing or its equivalent will be visible for inspection.
- Type AC cable shall have an armor of flexible metal tape and shall have an internal bonding strip of copper or aluminum in intimate contact with the armor for its entire length.
- May be installed exposed or concealed in dry locations.
- Not to be installed in damp or wet locations or where exposed to physical damage.
- Must be supported every 4½ ft (1.4 m) and within 12 in. (300 mm) of every box, fitting, or panel.

ARMORED CABLE RULES

Securing. Secured at intervals of 4½ ft (1.4 m), within 12 in. (300 mm) from every box, cabinet, or termination [320.30]. Staples or cable ties are most commonly used. Draping cable over air ducts, pipes, lower members of bar joists, and ceiling grid tees is not permitted.

Can be used unsupported where fished, in lengths of 2 ft (600 mm) when flexibility is needed, and in whips of 6 ft (1.8 m), connecting to equipment installed within accessible ceilings [320.30(D)].

Run horizontally through holes or notches in framing members spaced 4½ ft (1.4 m) apart, cable is adequately supported. Ties are not needed where the horizontal cable passes through these members. Cable must still be secured within 12 in. (300 mm) of each outlet.

Bending Radius. The radius of the curved inner edge of an AC cable bend must be five times the cable diameter [320.24],

ARMORED CABLE RULES

Thermal Insulation. Where more than two Type AC cables containing two or more current-carrying conductors in each cable are installed in contact with thermal insulation, caulk, or sealing foam without maintaining spacing between cables, the ampacity of each conductor shall be adjusted in accordance with Table 310.15(C)(1) [320.80(A)].

Protection Against Damage (Wood Framing). If run parallel to framing members, install so that the edge of the cable is 1¼ in. (32 mm) to any edge where screws or nails might penetrate. This applies to vertical framing members (e.g., studs) [300.4(D)].

Otherwise, protect cable with steel plate, sleeve, or other guard ¹⁄₁₆ in. (1.6 mm) thick.

Drill holes through vertical wood studs or other framing members at least 1¼ in. (32 mm) from the nearest edge or provide a steel protector [300.4(A)] for the AC cable.

Protection Against Damage (Metal Framing). Factory-made openings in steel studs keep cables 1¼ in. (32 mm) from the edge of the stud. Field-cut holes must comply with rules for holes drilled through wood framing members. Insulating bushings or grommets are not required when AC cable is run through openings in steel studs. The steel armor provides adequate protection against damaging the conductors inside.

In Exposed Locations. Cable must run along building finish surface or be on running boards. Exposed runs shall also be permitted to be installed on the underside of joists where supported at each joist and tocated so as not to be subject to physical damage [320.15].

* *Beneath joists.* Cables can be stapled directly to lower edges of joists, where they are supported at each joist and located to avoid damage [320.15].

* *Accessible attics.* Protect cables run atop floor joists, or across rafter or studding faces ≤ 7 ft (2.1 m) of the floor or floor joists, with guard strips for attics accessible by stairs. For a scuttle hole without stairs, guard strip protection is needed only within

⬛ ARMORED CABLE RULES

6 ft (1.8 m) of the hole's edge [320.23(A)]. Guard strips are not needed for cables run through holes bored in framing members, or parallel to sides of rafters, joists, and studs.

Cables run parallel to framing members or through bored holes must be 1¼ in. (32 mm) from the face of the stud or other framing member unless extra protection is provided [300.4(A)].

In Dropped or Suspended Ceilings. Properly secure and support as per 300.11(B).

Terminations. AC cables must terminate in compatible boxes or fittings. Install an insulating bushing at each termination: insert between the conductors and the armor for protection.

Connectors and clamps for use with AC cable are designed so that the insulating bushing is visible for inspection after installation.

Note: Metal Clad Cable, Type MC is a similar cable to Armored Cable: Type AC. Type MC cable is a factory assembly of one or more insulated circuit conductors with or without optical fiber members enclosed in an armor of interlocking metal tape, or a smooth or corrugated metallic sheath. See Article 330 for requirements pertaining to Type MC cable.

⬛ UNDERGROUND FEEDER AND BRANCH CIRCUIT CABLE: TYPE UF

Article 340 of *NEC*®

- Sizes 14 AWG copper through 4/0 AWG copper conductors.
- Cable contains equipment grounding conductor—may be bare or green.
- Cable may be buried in earth at minimum depth of 24 in.
- Burial depth may be reduced in some circumstances. See *NEC*® 300.5 and Table 300.5

SERVICE-ENTRANCE CABLE: TYPES SE AND USE

Article 338 of *NEC*®

- Service-entrance cable (SE) shall not be used under the following conditions or in the following locations:

  (1) Where subject to physical damage unless protected in accordance with 230.50(B)

  (2) Underground with or without a raceway

  (3) For exterior branch circuits and feeder wiring unless the installation complies with the provisions of Part I of Article 225 and is supported in accordance with 334.30 or is used as messenger-supported wiring as permitted in Part II of Article 396.

- Underground service-entrance cable (USE) shall not be used under the following conditions or in the following locations:

  (1) For interior wiring

  (2) For aboveground installations except where USE cable emerges from the ground and is terminated in an enclosure at an outdoor location and the cable is protected in accordance with 300.5(D)

  (3) As aerial cable unless it is a multiconductor cable identified for use aboveground and installed as messenger-supported wiring in accordance with 225.10 and Part II of Article 396

- Sometimes used for electric ranges and HVAC units.

- Cable contains equipment grounding conductor, which may be bare or green.

- Type USE may be buried in earth. See *NEC*® 300.5

BRANCH-CIRCUIT RULES

Code rules for grounded and grounding conductors in 120- and 240-volt branch circuits can be summarized as follows:

- Every 120-volt branch circuit consists of one grounded conductor, which has white or gray insulation as required by 200.6(A), and one ungrounded (hot or phase) conductor, which is usually black, though it may also be red or blue.
- Every 240-volt branch circuit consists of two ungrounded (hot or phase) conductors, which are usually black, red, or blue. Some 240-volt circuits have a grounded (white or gray) conductor.
- All 120-volt and 240-volt branch circuits have equipment grounding conductors. Sometimes these are called "green grounds" in the field, because many equipment grounding conductors have an outer covering that is predominantly green in color, as required by 250.119. However, 250.118 also permits bare conductors, metallic raceways, and the metal sheaths of some cables to be used for equipment grounding purposes.
- Grounded (white or gray) conductors and equipment grounding (green) conductors are not permitted to be connected to one another at any location except the service equipment.
- At panelboards other than service equipment, the equipment grounding conductor busbar is bonded to the metal enclosure or cabinet. But the neutral busbar is insulated or isolated to prevent electrical contact between it and the grounded metal panelboard enclosure. See requirements pertaining to objectionable current in 250.6.

Receptacle Ratings for Various Size Circuits

Circuit Rating (Amperes)	Receptacle Rating (Amperes)
15	Not over 15
20	15 or 20
30	30
40	40 or 50
50	50

Adapted from *NEC*® Table 210.21(B)(3).

BRANCH-CIRCUIT RULES

Summary of Branch-Circuit Requirement

Circuit Rating	15 A	20 A	30 A	40 A	50 A
Conductors (min. size):					
Circuit wires[1]	14	12	10	18	6
Taps	14	14	14	12	12
Fixture wires and cords— see 240.5					
Overcurrent Protection	**15 A**	**20 A**	**30 A**	**40 A**	**50 A**
Outlet devices:					
Lampholders permitted	Any type	Any type	Heavy duty	Heavy duty	Heavy duty
Receptacle rating	15 max. A	15 or 20 A	30 A	40 or 50 A	50 A
Maximum Load	**15 A**	**20 A**	**30 A**	**40 A**	**50 A**
Permissible load	See 210.23(A)	See 210.23(A)	See 210.23(B)	See 210.23(C)	See 210.23(C)

[1]These gauges are for copper conductors.
Adapted from *NEC*® Table 210.24.

⚡ VOLTAGE-DROP CALCULATION, 2-WIRE CIRCUITS

Use the following formula to calculate voltage drop in a two-wire alternating current (ac) circuit:

$$E_{VD} = IR$$

where:

E_{VD} = voltage drop

I = current in amperes (A)

R = resistance in ohms (Ω)

Use the formula to calculate the voltage drop over two 12 AWG solid copper conductors, each 85 ft (26 m) long, serving a 10-ampere load. Total circuit length is 170 tt (52 m).

NEC® Table 8: Conductor properties give different resistance values for coated copper and uncoated copper conductors. If the conductors are solid conductors, use 1.93 ohms per 1000 ft (305 m):

I = 10 amperes

R = 1.93 ohms per 1000 ft (305 m)

$$\frac{85 \times 2}{1000} = 0.170 \times 1.93$$

$$= 0.3281 \text{ ohms}$$

Substituting gives:
$$E_{VD} = IR$$
$$= 10 \text{ A} \times 0.3281 \ \Omega$$
$$= 3.281 \text{ volts (V)}$$

COMMON ELECTRICAL DISTRIBUTION SYSTEMS

120/240-Volt, Single-Phase, Three-Wire System

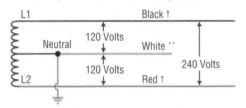

† • **Line one** ungrounded conductor colored **black**
† • **Line two** ungrounded conductor colored **red**
** • Grounded neutral conductor colored **white** or gray

120/240-Volt, Three-Phase, Four-Wire System (Delta High Leg)

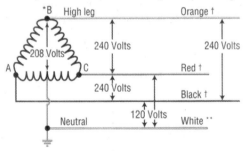

† • **A** phase ungrounded conductor colored **black**
†* • **B** phase ungrounded conductor colored **orange** or tagged
   (high leg). (Caution: 208 volts orange to white)
† • **C** phase ungrounded conductor colored **red**
** • Grounded conductor colored **white** or gray (center tap

**Grounded conductors are required to be white or gray or three white or gray stripes on other than green
   insulation. See NEC 200.6(A).
 * B phase of delta high leg must be orange or tagged. See *NEC* 110.15.
 † Ungrounded conductor colors may be other than shown; see local ordinances or specifications

120/208-Volt, Three-Phase, Four-Wire System (Wye Connected)

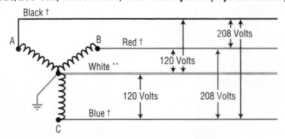

† • **A** phase ungrounded conductor colored **black**
† • **B** phase ungrounded conductor colored **red**
† • **C** phase ungrounded conductor colored **blue**
** • Grounded neutral conductor colored **white** or gray

277/480-Volt, Three-Phase, Four-Wire System (Wye Connected)

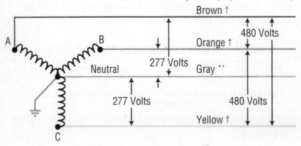

† • **A** phase ungrounded conductor colored **brown**
† • **B** phase ungrounded conductor colored **orange**
† • **C** phase ungrounded conductor colored **yellow**
** • Grounded neutral conductor colored **gray**

*Grounded conductors are required to be white or gray or three white or gray stripes on other than green insulation. See *NEC* 200.6(A).
† Ungrounded conductor colors may be other than shown; see local ordinances or specifications

🔌 WATER HEATER CIRCUIT CALCULATION

The nameplate rating for an electric water heater is 4500 watts (W), 240 volts (V), and the overcurrent rating for the circuit is not given:

$$\frac{4500 \text{ W}}{240 \text{ V}} = 18.75 \text{ A}$$

Section 422.11 (E)(3) requires overcurrent protection at not more than 150% of the rated current:

$$18.75 \text{ X } 150\% = 28.125 = 28 \text{ A}$$

The next higher standard overcurrent device rating is 30 amperes. See 240.6(A) and Table 240.6(A).

Therefore, a 30-ampere circuit breaker or fuse can be used to protect the electric water heater load in this example.

The branch-circuit rating for an appliance that is a continuous load, other than a motor-operated appliance, shall not be less than 125% of the marked rating [422.10(A)],

🔌 STANDARD AMPERE RATINGS FOR FUSES AND INVERSE TIME CIRCUIT BREAKER CIRCUITS

Table 240.6(A) Standard Ampere Ratings for Fuses and Inverse Time Circuit Breakers

Standard Ampere Ratings				
15	20	25	30	35
40	45	50	60	70
80	90	100	110	125
150	175	200	225	250
300	350	400	450	500
600	700	800	1000	1200
1600	2000	2500	3000	4000
5000	6000	—	—	—

Adapted from *NEC*® Table 240.6(A).

HVAC CIRCUITS

A Central Heating/Air-conditioning System with One Inside Unit (Furnace) and One Outside Unit (Compressor) Will Have Two Separate Branch Circuits

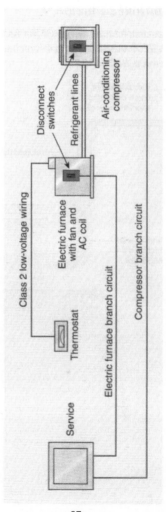

AIR-CONDITIONER WIRING

Overcurrent Protection for an Individual Motor Compressor Is Specified in *NEC*® 440.22(A). Branch-Circuit Conductor Ampacity Rules Are in *NEC* 440.32

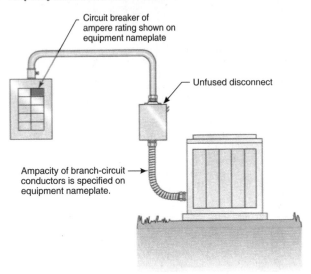

Circuit breaker of ampere rating shown on equipment nameplate

Unfused disconnect

Ampacity of branch-circuit conductors is specified on equipment nameplate.

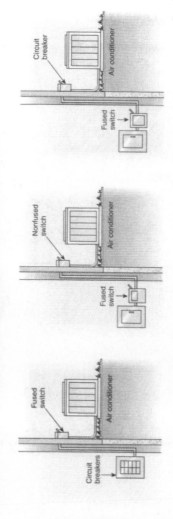

AIR CONDITIONER CONFIGURATIONS

These Three Wiring Configurations Can All be Used When the Unit Nameplate Specifies Fuse Protection

RESIDENTIAL LAMP TYPES

Lamps Used for Residential Lighting Come in Many Types and Shapes

Redrawn from *The IESNA Lighting Handbook,* 10th Edition, Courtesy of the Illuminating Engineering Society of North America

70

🔌 LIGHTING OUTLETS AND SWITCHES

Required Lighting Outlets

Section 210.70(A) requires the following locations in dwellings to have a lighting outlet:

- Every habitable room (living, dining, bedroom, den/library, family/rec, home office, etc.)
- Bathrooms
- Kitchens
- Hallways and stairways
- Attached garages
- Detached garages with electric power
- Attics, basements, utility rooms, and underfloor storage or equipment that requires servicing (e.g„ heating, air conditioning, and laundry; or water heaters)
- Dwellings or attached garage entrances with grade-level access (requires outdoor light controlled by in-house switch). Applies to entrances reached by stairs from grade level. A vehicle door in a garage is not considered an entrance [210.70(A) and (C)].

Switch Control of Lighting Outlets

- Each habitable room needs at least one wall switch-controlled lighting outlet [210.70(A)(1)],
- Interior stairways with six or more risers require a listed wall-mounted control device at each floor and landing level with an entryway [210.70(A)(2)(3)[. Local building codes may require a device for any stairway.
- Outdoor entrances to dwellings and garages with electric power must have a lighting outlet controlled by a listed wall-mounted control device [210.70(A)(2)(2)].

⚡ LIGHTING OUTLETS AND SWITCHES

- Lighting outlets can be controlled by occupancy sensors that are (1) in addition to listed wall-mounted control devices or (2) located at a customary wall switch location and equipped with a manual override that allows the sensor to function as a wall switch [210.70(A)(1) Exception No. 2],

- Exterior stairways are not required to be lighted. However, an outdoor entrance at the top of an exterior stairway must have a wall switch-controlled lighting outlet.

Switch Locations. *NEC*® does not specify where switches must be located. Exceed minimum *Code* requirements for greater convenience and homeowner satisfaction [90.1(B)],

- *Point of entry.* Most often, just inside a doorway on the side near the doorknob.

- *Multiway switching.* Install three- and four-way switches in rooms/hallways with multiple entrances. Multiway switches improve safety and convenience, and conserve energy by making it easier to turn off unneeded lighting.

- *Interior stairways.* Stairways with six or more risers require a listed wall-mounted control device at each floor level and landing with an entryway [210.70(A)(2)(3)].

- *Outside entrances.* Install wall switch inside each exterior door to control outside lighting outlet. (Note: Some builders install three-way switches that allow outdoor lighting to also be controlled from the master bedroom or other central location.)

- *Attached garages.* If a garage is attached to the house, it should be treated as another room for lighting and switches. Installing three- or four-way switches to control the lighting inside the garage is a convenient and safe solution. There should be a wall switch inside the garage vehicle door, another inside the exterior personnel door (if any), and one at the door leading into the house

LIGHTING OUTLETS AND SWITCHES

Switch-Controlled Receptacles

- In habitable rooms other than kitchens and bathrooms, instead of installing a lighting outlet, install receptacle outlet(s) controlled by a listed wall-mounted control device [210.70(A)(1), Exception No. 1].
- If a switch-controlled receptacle is used for lighting in a dining/breakfast room, or similar area, receptacle outlets on 20-ampere small-appliance branch circuits are still required [210.52(B)(1), Exception No. 1]. Small-appliance receptacle outlets may not be switched.
- Switch-controlled receptacle outlets are prohibited in hallways, stairways, garages, and at entrances. Switch-controlled lighting outlets are required [210.52(A)(2)],

For split-wired duplex receptacles, the top half is usually controlled by a switch; the bottom half is always live. A table lamp, floor lamp, or hanging lamp can then be plugged into a receptacle that is turned on and off, while the "always-on" receptacle is available for other loads.

To wire a duplex receptacle so that the top is switch-controlled and the bottom is always live:

1. Break off the tab joining the gold-colored screws on the receptacle. Do not break the tab joining the white-silver-colored screws.

2. Connect the receptacle branch-circuit conductors.

Switch and Receptacle Heights

Height is limited to 6 ft 7 in. (2.0 m) [404.8]. Typical ranges: 48-54 in. (1.2-1.4 m) for switches; 12-18 in. (305-457 mm) for receptacle outlets. Receptacles are usually mounted 6 in. (152 mm) above a counter surface or backsplash. Dimensions are measured to the device box centerline. Make heights of each device uniform throughout the dwelling.

The Americans with Disabilities Act Guidelines (ADAG) require wall switches in commercial construction be located up to 48 in. (1.2 m) AFF; receptacle outlets no more than 18 in. (457 mm) AFF. Follow ADAG when the house or apartment is intended for occupancy by persons with disabilities.

🔌 TRACK LIGHTING RULES

- *Mounting and supporting.* Lighting track shall be securely mounted so that each fastening is suitable tor supporting the maximum weight of luminaires that can be installed. Unless identified for supports at greater intervals, a single section 4 ft (1.2 m) or shorter in length shall have two supports, and, where installed in a continuous row, each individual section of not more than 4 ft (1.2 m) in length shall have one additional support [410.154], Lighting track is available for both surface and pendant mounting.

- *Electrical connections.* Either end-feed or center-feed; components are usually supplied to allow either type of connection. Connection must be grounded at outlet box. Track sections are required to be coupled securely enough to maintain grounding continuity [410.155(B)], Protect track section ends with insulating caps, supplied by manufacturer [410.155(A)],

- Connected load. The connected load on a lighting track shall not exceed the rating of the track. Lighting track shall be supplied by a branch circuit having a rating not more than that of the track [410.151 (B)].

Field modifications could damage track sections and continuous conductors, creating potential safety problems (poor fit, poor grounding, insecure power connections, accidental exposure of live conductors or energizing noncurrent-carrying parts of metal track).

Locations

Installed indoors, exposed, in dry locations; prohibited in the following locations [410.151(C)]:

- Where likely to be subjected to physical damage
- In damp or wet locations
- Where concealed
- Where extended through walls or partitions
- Less than 5 ft (1.5 m) above the finished floor, except where protected from physical damage or track operating at less than 30 volts rms open-circuit voltage
- In a zone extending 3 ft (900 mm) horizontally and 8 ft (2.4 m) vertically from the top of a bathtub rim or shower threshold

🔌 BASEMENT LIGHTING GUIDELINES

- *Usual point of entry.* At least one switch should be provided at the "usual point of entry" to a basement or utility room, as required by 210.70(A)(1). If there are two or more points of entry, multiple switches are needed.

- *Basement stairs.* When a basement is entered by a stairway from the floor above, the stairway itself becomes the point of entry and requires at least one light controlled by a switch. In this case, the stairway lighting can be controlled from the top of the stairs, and the lighting within the basement can be controlled by its own switch. However, the preferable method is to install three-way switches that allow control from both the top and foot of the stairs.

- *Multiway switching.* Using three- and four-way switches is often the most convenient way to control basement lighting when there is more than one entrance.

🔌 HALLWAY LIGHTING GUIDELINES

- *Single point of entry.* A hallway with only one point of entry (such as one leading to a closet or powder room) should have a wall switch placed where a person enters the hallway. Only one switch is needed for this type of hallway.

- *Two or more points of entry.* A hallway that can be entered from more than one point should use multiway switching (three- or four-way switching). The goal is to avoid forcing homeowners to walk through a dark space looking for a light switch—that is how accidents happen. As a rule of thumb, residents should not have to walk more than 6 ft (1.8 m) to find a hallway lighting switch.

- *Hallway leading to habitable room(s).* A hallway leading to habitable rooms should also use multiway switches.

CALCULATING COST OF OPERATING AN ELECTRICAL APPLIANCE

What is the monthly cost of operating a 240-volt, 5-kilowatt (kW) central electric heater that operates 12 hours per day when the cost is 15 cents per kilowatt-hour (kWhr)?

Cost = Watts x Hours Used x Rate per kWhr/1000

5 kW = 5000 watts

hours = 12 hours x 30 days = 360 hours per month

= 5000 x 360 x 0.15/1000

= 270000/1000 = **$270 Monthly Cost**

The above example is for a resistive load. Air-conditioning loads are primarily inductive loads. However, if ampere and voltage values are known, this method will give an approximate cost. Kilowatt-hour rates vary for different power companies, and for residential use, graduated-rate scales are usually used (the more power used, the lower the rate). Commercial and industrial rates are generally based on kilowatt usage, maximum demand, and power factor. Other costs are often added, such as fuel cost adjustments.

 CHANGING INCANDESCENT LAMP TO ENERGY-SAVING LAMP

A 100-watt incandescent lamp is to be replaced with a 15-watt, energy-saving lamp that has the same light output (lumens). If the cost per kilowatt-hour (kWhr) is 15 cents, how many hours would the new lamp need to operate to pay for itself?

tamp cost is 4 dollars. Energy saved is 85 watts.

Hours = Lamp Cost x 1000/Watts Saved x kWhr

(4 x 1000)/(85 x 0.15) = 4000/12.75 = 313.73 hours

The energy-saving lamp will pay for itself with 313.73 hours of operation. The comparative operating cost of these two lamps based on 313.73 hours is found by:

Cost = Watts x Hours Used x Rate per kWhr/1000

100-watt incandescent lamp = $4.71 for 313.73 hours of operation

15-watt energy saving lamp = $0.71 for 313.73 hours of operation

SINGLE-POLE, THREE- AND FOUR-WAY SWITCHING

Three-Way Switches Are Used to Control a Load from Two Different Locations

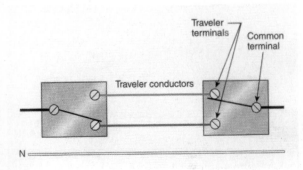

⚡ SINGLE-POLE, THREE- AND FOUR-WAY SWITCHING

A Single-Pole Switch Controls a Lighting Outlet, with Power Feed at the Outlet

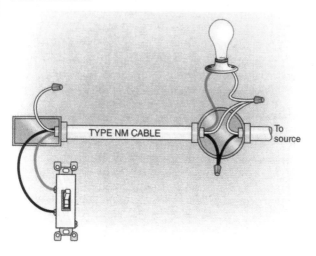

TYPE NM CABLE

To source

⌫ SINGLE-POLE, THREE- AND FOUR-WAY SWITCHING

A Single-Pole Switch Controls a Lighting Outlet, with Power Feed at the Switch

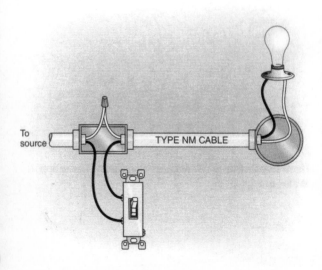

To source

TYPE NM CABLE

SINGLE-POLE, THREE- AND FOUR-WAY
SWITCHING

Three-Way Switching Can Be Installed with the Power Supply at the First Switch

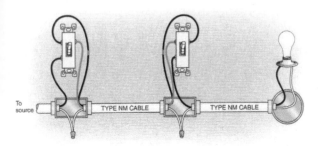

To source — TYPE NM CABLE — TYPE NM CABLE

Three-Way Switching Can Also Operate with the Power Supply at the Load

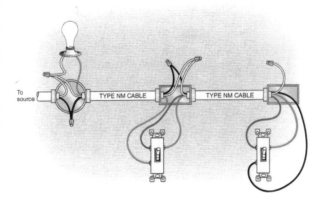

To source — TYPE NM CABLE — TYPE NM CABLE

SINGLE-POLE, THREE- AND FOUR-WAY SWITCHING

Three-Way Switching Can Be Wired as Shown with the Power Supply at the Load between the Two Switches

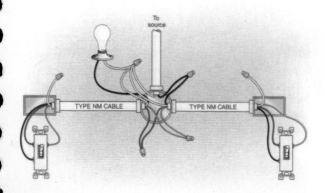

Four-Way Switching, with the Power Supply at the Last Switch, Allows Control from Three Locations

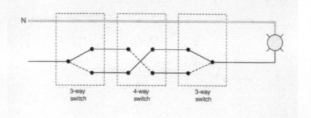

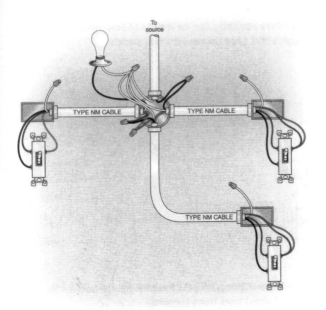

RECEPTACLE LAYOUTS

This Living Room Layout Uses a Floor-Mounted Receptacle Outlet (Six Receptacles Total)

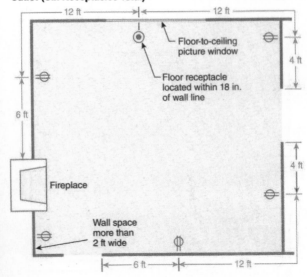

Floor-to-ceiling picture window

Floor receptacle located within 18 in. of wall line

Fireplace

Wall space more than 2 ft wide

RECEPTACLE LAYOUTS

This Living Room Receptacle Layout Has No Floor-Mounted Receptacle Outlet (Eight Receptacles Total)

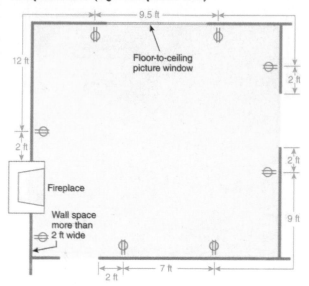

84

BATHROOM WIRING

The Bathroom Has One Basin Located Outside the Door to the Rest of the Bathroom Area

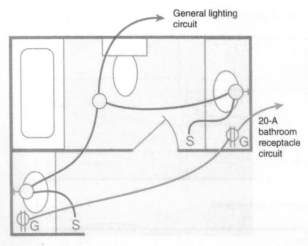

General lighting circuit

20-A bathroom receptacle circuit

G = GFCI protection required

BATHROOM WIRING

Two Different Arrangements of GFCI-Protected Receptacles Can Be Used in this Bathroom

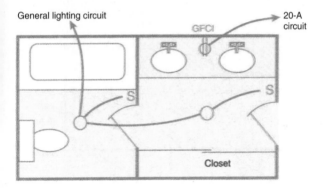

General lighting circuit

GFCI

20-A circuit

S

S

Closet

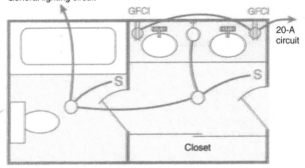

General lighting circuit

GFCI

GFCI

20-A circuit

S

S

Closet

BATHROOM WIRING

A Receptacle in the Location Shown Does Not Satisfy
***NEC*® 210.52(D) Because It Is Not "On a Wall or Partition That Is Adjacent to the Basin or Basin Countertop"**

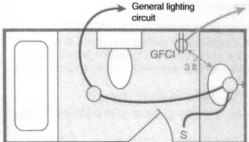

General lighting circuit

20-A bathroom receptacle circuit

GFCI

3 ft

S

⚡ REQUIRED GFCI LOCATIONS

All 125-volt through 250-volt receptacles installed in the locations specified in 210.8(A)(1) through (11) and supplied by single-phase branch circuits rated 150 volts or less to ground shall have ground-fault circuit-interrupter protection for personnel. GFCI protection is required in the following dwelling unit areas:

- Bathrooms, at least one required. See 210.8(A)(1), 210.52(D).

- Garages, and also accessory buildings that have a floor located at or below grade level not intended as habitable rooms and limited to storage areas, work areas, and areas of similar use. See 210.8(A)(2), 210.52(G).

- Outdoors. See 210.8(A)(3), 210.8(F), 210.52(E).

- Crawl spaces—at or below grade level. See 210.8(A)(4), 210.8(E), 210.63.

- Basements. See 210.8(A)(5), 210.52(G).

- Kitchens—where the receptacles are installed to serve the countertop surfaces. See 210.8(A)(6), 210.52(C).

- Sinks—where receptacles are installed within 6 ft (1.8 m) from the top inside edge of the bowl of the sink. See 210.8(A)(7), 210.52(B) and (C).

- In boathouses and for boat hoists. See 210.8(A)(8), 210.8(C).

- Bathtubs or shower stalls—where receptacles are installed within 6 ft (1.8 m) of the outside edge of the bathtub or shower stall. See 210.8(A)(9), 210.52(D).

- Laundry areas. See 210.8(A)(10), 210.52(F).

- Indoor damp and wet locations. See 210.8(A)(11).

- Dishwashers and sump pumps. See 422.5(A)(6) and (7).

- Receptacles and equipment around pools and spas. See Article 680. Some of the requirements are in 680.5, 680.21 (C), 680.22(A)(4), 680.22(B)(4), 680.23(A)(3), 680.32, 680.43(A)(2), 680.44, 680.45, 680.51, 680.62, and 680.71.

- Lighting outlets not exceeding 120 volts in crawl spaces. See 210.8(C).

AFCI AND TAMPER-RESISTANT RECEPTACLES

Dwelling units are required to have arc-fault circuit-interrupter (AFCI) protection for all 120-volt, single-phase, 15- and 20-ampere branch circuits supplying outlets or devices installed in kitchens, family rooms, dining rooms, living rooms, parlors, dens, libraries, bedrooms, sunrooms, recreation rooms, closets, hallways, laundry areas, or similar rooms or areas. See *NEC®* 210.12 for more details.

Dwelling units are required to have tamper-resistant receptacles installed in kitchens, family rooms, living rooms, parlors, libraries, dens, sunrooms, recreation rooms, bathrooms, laundry rooms, basements garages, hallway, bedrooms, and outdoors. See NEC® 210.52 and 406.12 for more details.

POOL AREA RECEPTACLES

Receptacle Outlets Around Dwelling-Unit Swimming Pools

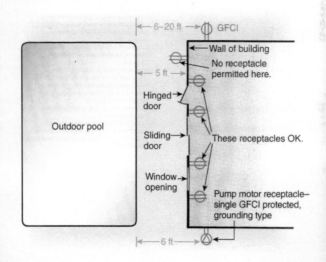

⊡ POOL LIGHTING

Luminaires, Lighting Outlets, and Ceiling-Suspended (Paddle) Fans Cannot Be Installed in the Areas Shown that Surround Outdoor and Indoor Pools

Outdoor Pools

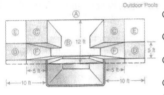

Ⓐ Luminaires, lighting outlets, and ceiling-suspended (paddle) fans permitted above 12ft.

Ⓑ Luminaires, lighting outlets, and ceiling-suspended (paddle) fans not permitted below 12 ft.

Ⓒ Existing luminaires and lighting outlets permitted in this space if rigidly attached to existing structure (GFCI required).

Ⓓ Luminaires and lighting outlets permitted if protected by a GFCI.

Ⓔ Luminaires and lighting outlets permitted if rigidly attached.

Ⓕ Listed low-voltage luminaires not requiring grounding and not exceeding the low-voltage contact limit, powered by supplies in accordance with 680.23(A)(2).

Indoor Pools

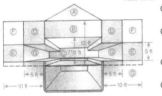

Ⓐ Luminaires, lighting outlets, and ceiling-suspended (paddle) fans permitted above 12 ft.

Ⓑ Totally enclosed luminaires protected by a GFCI and ceiling-suspended (paddle) fans protected by a GFCI permitted above 7½ ft.

Ⓒ Luminaires, lighting outlets, and ceiling-suspended (paddle) fans not permitted below 5 ft.

Ⓓ Existing luminaires and lighting outlets permitted in this space if rigidly attached to an existing structure (GFCI required).

Ⓔ Luminaires and lighting outlets permitted if protected by a GFCI.

Ⓕ Luminaires and lighting outlets permitted if rigidly attached.

Ⓖ Listed low-voltage luminaires not requiring grounding and not exceeding the low-voltage contact limit, powered by supplies in accordance with 680.23(A)(2

90

POOL AREA CLEARANCES—OVERHEAD CONDUCTORS

Clearance Details	Insulated Cables (0-750 Volts to Ground), Supported on and Cabled with a Solidly Grounded Bare Messenger or Neutral Conductor	All Other Conductors Voltage to Ground	
		0–15 kV	Over 15–50 kV
	ft	ft	ft
A. Clearance (in any direction) to edge of water surface, base of diving platform, or to the water level	22.5	25	27
B. Clearance (in any direction) to tower, observation stand, or to diving platform	14.5	17	18
C. Clearance has horizontal limit measured from inside wall of pool	Horizontal limit shall extend to outside edge of structures indicated in A and B above but not less than 10 feet.		

Adapted from *NEC*® Table 680.9(A).

RECOMMENDED HEIGHTS FOR WALL-MOUNTED OUTLETS

Outlet	Height to Center (in.)
Switches, dimmers, fan controls	48
Receptacle outlets, general	18
Receptacle outlets, kitchen, utility room, workbenches, etc.	42, or 6 above countertop or backsplash
Telephones, general	18
Intercom/music station	48
Telephones, wall mounted	60
Thermostats	60
Security system keypads	48
Kitchen clock hangers	6 below ceiling; above doors, center the clock outlet between door trim and ceiling
Doorbells, buzzers, chimes	78
Lighting	78, typical; varies depending on size and design of luminaire (lighting fixture)

CONDUCTORS FOR SINGLE-PHASE DWELLING SERVICES AND FEEDERS

Conductor Types/Sizes tor 120/240-Volt, 3-Wire, Single-Phase Dwelling Services and Feeders

Service or Feeder Rating (Amperes)	Conductor (AWG or kcmil	
	Copper	Aluminum or Copper-Clad Aluminum
100	4	2
110	3	1
125	2	1/0
150	1	2/0
175	1/0	3/0
200	2/0	4/0
225	3/0	250
250	4/0	300
300	250	350
350	350	500
400	400	600

For one-family dwellings and the individual dwelling units of two-family and multifamily dwellings, service and feeder conductors supplied by a single-phase, 120/240-volt system shall be permitted to be sized in accordance with 310.12(A) through (D).
Adapted from *NEC®* 310.12

SINGLE-FAMILY DWELLING—SERVICES

Service Point Location

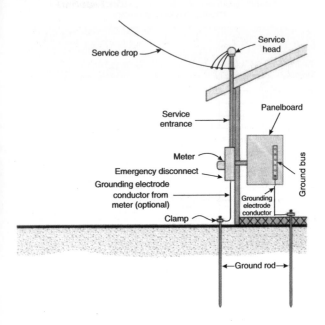

Service drop

Service head

Service entrance

Panelboard

Meter

Emergency disconnect

Grounding electrode conductor from meter (optional)

Clamp

Grounding electrode conductor

Ground bus

Ground rod

Service Entrance Bonding

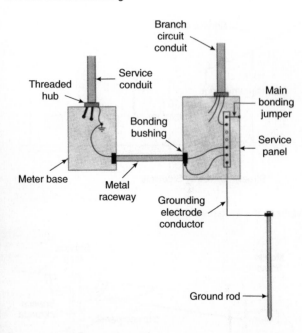

Branch circuit conduit

Service conduit

Threaded hub

Main bonding jumper

Bonding bushing

Service panel

Meter base

Metal raceway

Grounding electrode conductor

Ground rod

SINGLE-FAMILY DWELLING—SERVICES

Service-Drop Clearance

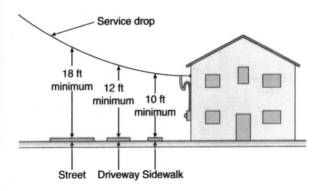

Service drop

18 ft minimum

12 ft minimum

10 ft minimum

Street | Driveway | Sidewalk

Overhead Mobile Home Service

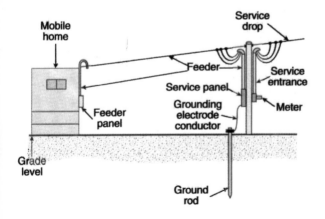

Mobile home

Service drop

Feeder

Service entrance

Service panel

Feeder panel

Grounding electrode conductor

Meter

Grade level

Ground rod

SINGLE-FAMILY DWELLING—SERVICES

Underground Mobile Home Service

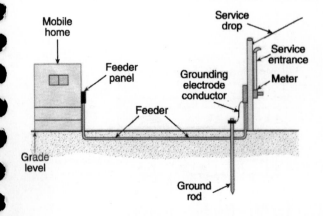

MULTI-FAMILY SERVICES

**This Illustration Shows Meters and Service Equipment for a
Two-Family Dwelling with an Overhead Service**

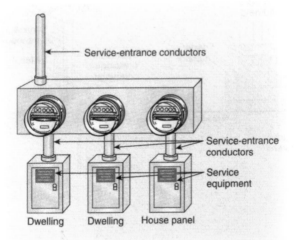

Service-entrance conductors

Service-entrance conductors

Service equipment

Dwelling   Dwelling   House panel

🔌 MULTI-FAMILY SERVICES

This Illustration Shows Meters and Panelboards for a Two-Family Dwelling with an Underground Service. The Panelboards Are Not Required to Be Listed as Service Equipment

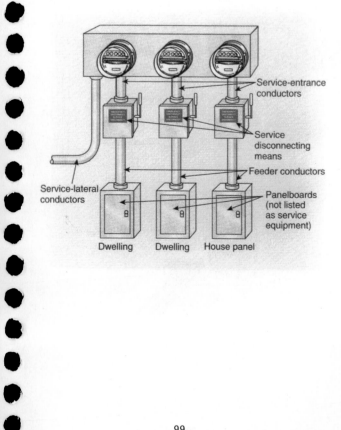

Service-entrance conductors

Service disconnecting means

Feeder conductors

Service-lateral conductors

Panelboards (not listed as service equipment)

Dwelling   Dwelling   House panel

A Meter Pedestal Can Be Used with Underground Service for a Two-Family Dwelling

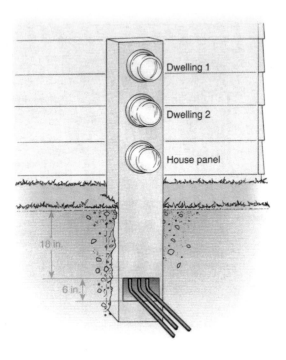

Dwelling 1

Dwelling 2

House panel

18 in.

6 in.

MULTI-FAMILY SERVICES

This Illustration Shows Grounding and Bonding Arrangements for Three Switches That Are the Disconnecting Means for a Single Service to a Two-Family Dwelling

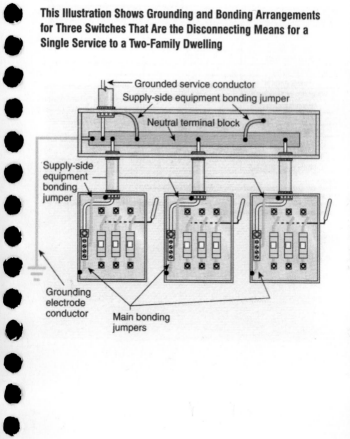

Grounded service conductor
Supply-side equipment bonding jumper
Neutral terminal block
Supply-side equipment bonding jumper
Grounding electrode conductor
Main bonding jumpers

TYPES OF APPLIANCE BRANCH CIRCUITS

Type	Voltage	Pole Spaces
Range (electric)	250	2
Range (gas)	*	*
Oven (electric)	240	2
Oven (gas)	*	*
Refrigerator†	120	1
Water heater (electric)	240	2
Water heater (gas)	120	1
Furnace (electric)	240	2
Furnace (gas, oil)	120	1
Dishwasher/disposal	120	1
Clothes dryer (electric)	240	2
Clothes dryer (gas)	**	**

Notes:
*Supplied by small-appliance branch circuits [210.52(B)(2), Exception No. 2].
**Supplied by laundry branch circuits [210.11(C)(2)].
†The receptacle outlet for a refrigerator may be supplied by the small-appliance branch circuits [210.52(B)(1)] or by a separate circuit [210.52(B)(1), Exception No. 2].

⚡ APPLYING THE STANDARD LOAD CALCULATIONS

The following loads are separated into two columns of loads, and demand factors are applied.

Column 1 is the general lighting and receptacle loads and small appliance loads **Table 220.42.**

Column 2 is the cooking equipment **loads 220.55** and **Table 220.55,** fixed appliance loads **220.53, and** dryer loads **220.54** and **Table 220.54.**

Column 1: General Lighting Load—220.14(J).

Multiply the building square footage by 3 VA per 220.14(J).

A minimum of two small-appliance circuits are required to supply receptacles located in the kitchen, breakfast room, pantry, and dining room per *NEC®* **220.52(A).** An additional small-appliance load is required to service the laundry loads per *NEC®* **220.52(B).** The small-appliance and laundry circuits shall be calculated at 1500 VA each.

After the General Lighting Load is totaled, the demand factor per *NEC®* Table 220.42 is applied per the occupancy type. For dwellings units, the first 3000 VA shall be calculated at 100%. From 3001 to 120000 VA shall be calculated at 35%, and the remaining at 25%. See the Column 1 breakdown below.

Column 1: General Lighting Load—220.14(J)

_____ sq. ft x 3 VA   = _____VA

Small-Appliance Loads—220.52(A) and (B)

_____VA X _____	= _____VA	Ungrounded	Grounded
Total	= _____VA	(phase) Load	(neutral) Load

General Lighting Load and Demand Load 1—Table 220.42

First 3000 VA @ 100%	= _____ VA	
Next _____VA @_____%	= _____ VA	
Remainder _____VA @_____%	= _____ VA	
Total	= _____ VA _____ **VA**	_____**VA**

APPLYING THE STANDARD LOAD CALCULATIONS

Column 2: Special Appliance Loads

220.14(A) thru (C) and 230.42

Usually special appliance loads are supplied by direct circuits. These may be loads like water heaters, heating units, ranges, air conditioners, dishwashers, motors, and so on. Direct circuit loads are not connected to the general lighting load. The overcurrent protection device VA rating shall be 125% for continuous operation and 100% for noncontinuous loads with demand loads being determined by the nameplate kilowatt rating times a percentage.

The total number of fixed-appliance loads shall have demand factors applied. The total number of ranges and dryers in a dwelling unit shall be reduced by a percentage.

Column 2: Special Appliance Loads—Table 220.55, Table 220.54, and 220.53 Cooking Load and Demand Load 2—Table 220.55

Table 220.55	Phases	**220.61**	Neutral	
Column C	____VA × 70%		= ____VA	
Table 220.55	Total VA	**Table 220.55**	Phases	
Column B	____VA × ____%		= ____VA	
	Phases	**220.61**	Neutral	
	____VA × 70%		= ____VA ____VA ____VA	

Dryer Loads and Demand Load 3—Table 220.54

Table 220.54	Phases	**Table 220.54**	Phases	____VA ____VA
	____VA	× ____%	= ____VA	
220.61	Phases	**220.61**	Neutral	All loads / 120-volt load
	____VA × 70%		= ____VA	**220.53** / **220.61**

Fixed-Appliance Load and Demand Load 4—220.53
(Calculate all loads and 120-volt by 75%)   ____VA ____VA

____VA + ____VA + ____VA + ____VA × 75%	= ____VA	(120V, 1Ø)
•____VA + •____VA + •____VA + •____VA × 75%	= ____VA	
Total	= ____VA	(120/240V, 1Ø)

Note: • = 240-volt loads

104

APPLYING THE STANDARD LOAD CALCULATIONS

Cooking Equipment Loads—Column 2—220.55 and Table 220.55

Cooking equipment loads and their demand load two are placed in Column 2 with demand factors applied. The demand factors are listed in Table 220.55 (see page 111).

Fixed Appliance Load—Column 2—220.53

Fixed-appliance load for three or less fixed appliances shall be determined by adding wattage (volt-amps) values by the nameplate ratings for each appliance.

Fixed-appliance loads for four or more fixed appliances shall be determined by adding wattage (volt-amps) values by the nameplate rating for each appliance with a 75% demand factor applied per **220.53**.

Dryer Load—Column 2—220.54 and Table 220.54

Demand loads for household dryers are calculated at the larger of 5kVA or the nameplate rating. Four or less dryers shall be calculated at 100% of the nameplate rating. Five or more dryers shall have a percentage applied based on the number of units per Table **220.54**.

Household Dryers—Demand Factors

Demand Factor (%)	Number of Dryers
100	1-4
85	5
75	6
65	7
60	8

Adapted from NEC® Table 220.54.

APPLYING THE OPTIONAL LOAD CALCULATIONS

Largest Load Between Heating and Air Conditioning—Column 3—220.60

Per section **220.60**, the smaller of the two loads will be dropped between the heating and the air-conditioning loads. The remaining loads shall be calculated at 100%. Where a motor is part of a noncoincident load and is not the largest of the noncoincident loads, 125% of the motor load shall be used in the calculation if it is the largest motor.

Largest Motor Load—Column 4—220.50

The largest motor full load current rating shall be calculated at 125% per **220.50**. See **430.24** and **430.25** for loads of one or more motors with other loads.

The largest motor load shall always be added into the calculation.

Column 3—Heating or A/C Load—220.60	____VA	____VA
Column 4—Largest Motor Load—220.50	____VA	____VA
____VA × 25% = ____VA **TOTAL VA**	**VA**	**VA**
• ____VA × 25% = ____VA		

Ungrounded (phase) conductors	Grounded (neutral) conductor
I = ____VA ÷ ____ V = ____ A	I = ____VA ÷ ____ V = ____ A

Note: • = 240-volt loads

Refer to *NEC*® **Annex D, Example D1** for an example of Standard Load Calculation Method.

🔌 APPLYING THE OPTIONAL LOAD CALCULATIONS

The optional calculation is an easier method for calculating the load per **220.82(B)** and **(C)**. The loads are separated into two columns. The first column of loads consists of all nonheating and air-conditioning loads. The heating and air-conditioning loads are in the second column. This method has percentages applied that are derived from the demand factors. The optional method can only be used when an ampacity of 100 amps or greater is applied to the service conductors.

General Loads—220.82

The first column is the general loads. General lighting and general-purpose receptacle loads are calculated at 3 VA per sq. ft per **220.82(B)(1)**. Small appliance loads are calculated at 1500 VA per 220.82(B)(2) as covered in **210.11(C)(1)** and **(C)(2)**. The special appliance loads are added with the general loads per **220.82(B)(3)**. Section **220.82(B)(4)** includes motor loads in the general loads column. The largest motor does not have to be calculated at 125%. The first 10,000 VA shall be calculated at 100%. The remaining VA shall be calculated per **220.82(B)** at 40%.

Heat or Air-Conditioning Loads—220.82(C)(1) thru (C)(6)

The second column is the heating, air-conditioning, or heat pumps loads. Three or fewer heating units shall be calculated at 65% of the total kW. Four or more heating units shall be calculated at 40% of the total kW. The air-conditioning loads are calculated at 100% of the kVA rating. The total heating load and the air-conditioning loads are compared, and the smaller load is discarded.

Refer to **NEC® Annex D, Example D2** for an example of Optional Load Calculation Method.

APPLYING THE OPTIONAL LOAD CALCULATIONS

Column 1—Other Loads—220.82(B)		Ungrounded (phase) Load	Grounded (neutral) Load
General Lighting Load—220.82(B)(1)			
___ sq. ft x ___ VA = ___ VA		___ **VA**	
Small Appliance Load—220.82(B)(2)			
___ VA x ___ = ___ VA		___ **VA**	
General lighting load	Demand load 1 N		___ **VA**
Appliance Load—220.82(B)(3) and (B)(4)			
Range load • (Cooking equipment load)	Demand load 2 N	___ **VA**	___ **VA**
Cooktop load •		___ **VA**	
Oven load •		___ **VA**	
Dryer load • = (Dryer equipment)	Demand load 3 N	___ **VA**	___ **VA**
Water heater load • = (Fixed-appliance load)	Demand load 4 N	___ **VA**	___ **VA**
Disposal load = (Heating or A/C load)	Remaining load 5 N	___ **VA**	___ **VA**
Compactor load = (Largest motor load)	Remaining load 6 N	___ **VA**	___ **VA**
Dishwasher load		___ **VA**	
Attic fan load		___ **VA**	
Water pump load		___ **VA**	
Blower motor load		___ **VA**	
Sump pump load		___ **VA**	
Pool pump load		___ **VA**	
Water pump load		___ **VA**	
___ load		___ **VA**	
Total load		___ **VA**	___ **VA**

Demand Load—220.82(B)		Add Columns 1 and 2 to derive total service load in VA
First 10000 VA @ 100%	= 10000 VA	
Remainder ___ VA @ 40%	= ___ VA	
Total load	= ___ VA	___ **VA** (Column 1)

Column 2—Largest Load Between Heating and A/C Load
220.82(C)(1), (C)(2), and (C)(4)

Heating load	___ VA × # ___ × ___ % =	___ **VA**	(Column 2)
A/C load	___ VA × # ___ × 100% =	___ **VA**	(Column 2)
Heating pump load	___ VA × # ___ × 100% =	___ **VA**	(Column 2)
Total load	=	___ **VA**	

Ungrounded (phase) conductors
I = ___ VA ÷ ___ V = ___ A

Grounded (neutral) conductor
I = ___ VA ÷ ___ V = ___ A

Note: • = 240-volt loads

NEC®220.14(J) specifies a lighting load of three volt-amperes per square foot for dwelling units, and 210.11(B) requires that this load be evenly proportioned among multioutlet branch circuits. Assuming a house floor area of 2200 ft² (204 m²), the minimum number of lighting branch circuits needed is calculated using the following formula:

$$\frac{3 \text{ volt-amperes (VA) x sq. ft}}{120 \text{ volts}} = amperes$$

which gives

$$\frac{3 \text{ VA x 2200 sq. ft}}{120 \text{ V}} = 55 \text{ A}$$

Using 15-ampere branch circuits,

$$\frac{55 \text{ A}}{15 \text{ A}} = 3.67 \text{ branch circuits (round up to 4 circuits)}$$

This result works out to one 15-ampere lighting branch circuit per each 550 ft² or 51 m² (2200 ÷ 4 = 550).

Using 20-ampere branch circuits,

$$\frac{55 \text{ A}}{20 \text{ A}} = 2.75 \text{ branch circuits (round up to 3 circuits)}$$

This result works out to one 15-ampere lighting branch circuit per each 733 ft² or 68 m² (2200 ÷ 3 = 733).

APPLIANCE LOAD CALCULATION

Determine the 120/240-volt service load needed for the following fastened-in-place appliances in a dwelling unit

Appliance	Rating	Load
Water heater	4500 W, 240 V	4500 VA
Kitchen waste disposal	¼ hp, 120 V	696 VA
Dishwasher	1250 W, 120 V	1250 VA
Furnace motor	¼ hp, 120 V	696 VA
Whole-house fan	½ hp, 120 V	1176 VA

First, calculate the total of the five fastened-in-place appliances (in volt-amperes):

Total load = 4500 VA + 696 VA + 1250 VA + 696 VA + 1176 VA
= 8318 VA

Because the load is for more than four appliances, apply a demand factor of 75%:

$$8318 \text{ VA} \times 0.75 = 6239 \text{ VA}$$

DEMAND FACTORS FOR RANGES AND OTHER COOKING APPLIANCES OVER 1¾ kW RATING

(Column C to be used in all cases except as otherwise permitted in Note 3. See next page.)

Number of Appliances	Demand Factor (%) (See notes on the following page)		
	A (Less than 3½ kW Rating)	B (3½ kW Through 8¾ kW Rating)	C maximum Demand (kW) (Not over 12 kW Rating)
1	80	80	8
2	75	65	11
3	70	55	14
4	66	50	17
5	62	45	20
6	59	43	21
7	56	40	22
8	53	36	23
9	51	35	24
10	49	34	25
11	47	32	26
12	45	32	27
13	43	32	28
14	41	32	29
15	40	32	30
16	39	28	31
17	38	28	32
18	37	28	33
19	36	28	34
20	35	28	35
21	34	26	36
22	33	26	37
23	32	26	38
24	31	26	39
25	30	26	40
26-30	30	24	15 kW + 1 kW for each range
31-40	30	22	
41-50	30	20	25 kW + ¾ kW for each range
51-60	30	18 >	
61 and over	30	16	

(continued on next page)

DEMAND FACTORS FOR RANGES AND OTHER COOKING APPLIANCES OVER 1¾ kW RATING

Notes:
1. Over 12 through 27 kW range: For ranges individually rated in this range, the maximum demand in Column C is increased 5% for each additional kilowatt exceeding 12 kW.
2. Over 8¾ through 27 kW range: For ranges individually rated more than 83A kW and of different ratings, not exceeding 27 kW, the average value shall be calculated (using 12 kW for any range less than 12 kW). Maximum demand in Column C is then increased 5% for each additional kilowatt exceeding 12 kW.
3. Over 1¾ through 8¾ kW range: Instead of the method provided in Column C, it is permissible to add the nameplate ratings for all household cooking appliances rated in this range and multiply the sum by the demand factors as indicated in Columns A or B for the number of appliances. If an appliance falls into both columns, the demand facts for each column shall be applied to the appliances for that column and the results added together.
4. It is permitted to calculate the branch-circuit load for a range per Table 220.55. The branch-circuit load for a wall-mounted oven or counter-mounted cooking unit shall be per the nameplate rating of the appliance. The branch-circuit load for a counter-mounted cooking unit and up to two wall-mounted ovens, all fed from a single branch circuit and in the same room, shall be calculated by adding the nameplate ratings and treating this total as equal to one range.
5. This table shall also apply to instructional programs using household cooking appliances rated over 13A kW. Adapted from NEC® Table 220.55

 NEUTRAL REDUCTION RULES

The neutral conductor of a 120/240-volt, single-phase, 3-wire branch circuit supplying an electric range, wall-mounted oven, or cooktop can be smaller than the ungrounded (phase) conductors when the maximum demand is 8¾ kW, as computed according to *NEC®* Table 220.55, Column C. The neutral ampacity must be at least 70% of the branch-circuit rating, and the conductor must be at least 10 AWG [210.19(A)(3), Exception No. 2],

Column C of Table 220.55 indicates that the maximum demand for one range rated 12 kW or less is 8 kW:

$$8 \text{ kW} = 8000 \text{ VA} - 240 \text{ V} = 33.3 \text{ A}$$

Thus, the ungrounded range circuit conductors can be 8 AWG, Type TW copper, rated 40 amperes.

The neutral of this 3-wire circuit is permitted to be 10 AWG, which is smaller than 8 AWG but which has an ampacity (30 amperes) that is more than 70% of the 40-ampere rating of the phase conductors. The maximum neutral demand of a range circuit seldom exceeds 25 amperes. Current is drawn from the neutral only to power 120-volt lights, clocks, timers, and sometimes heating elements in their low-heat positions.

RESIDENTIAL NEUTRALS

Neutrals Are One Type of Grounded Conductor

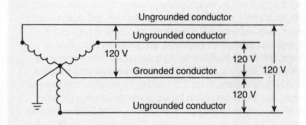

In a Balanced 4-Wire Circuit, the Grounded Conductor Is Neutral.

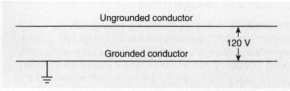

In a 2-Wire Circuit, the Grounded Conductor Is Not Truly
Neutral, but Is a Current-Carrying Conductor.

PARTIAL 2020 NATIONAL ELECTRICAL CODE SUMMARY FOR RESIDENTIAL WIRING

Structural Overhangs [110.26(E)(2)(c) Exception]

Since the 2014 edition, dedicated equipment space has been required for outdoor switchboards, switchgear, panelboards, and motor control centers. Required dedicated equipment space is the space equal to the width and depth of the equipment, and extending from grade to a height of 6 ft (1.8 m) above the equipment. Especially in residential installations, structural overhangs and roof extensions are sometimes in that dedicated space. In the 2017 edition, an exception was added to address this type of installation but the exception was inserted under 110.26(E)(2)(b), which pertains to workspace. In the 2020 edition, the location of the exception was corrected. The exception under dedicated equipment space says structural overhangs or roof extensions shall be permitted in this zone.

GFCI Protection for Personnel [210.8]

When distance is a factor when determining if ground-fault circuit-interrupter (GFCI) protection is required for receptacles, passing through a door or doorway is no longer a reason for not providing GFCI protection for those receptacles. For example, if a receptacle is located in a cabinet under a kitchen sink and that receptacle is within 6 ft (1.8 m) of the sink, that receptacle is required to be GFCI protected. Also, if a receptacle in a bedroom (or hallway) is within 6-ft of a sink in the bathroom as measured through the bathroom doorway, that bedroom (or hallway) receptacle is required to be GFCI protected.

GFCI Protection in Dwellings [210.8(A)]

Ground-fault circuit-interrupter (GFCI) protection for personnel is also now required for all 250-volt receptacles installed in the locations specified in 210.8(A)(1) through (A)(11) and supplied by single-phase branch circuits rated 150 volts or less to ground.

GFCI Protection in Basements [210.8(A)(5)]

This section now applies to all dwelling basements, not just unfinished portions or areas of the basement not intended as habitable rooms.

PARTIAL 2020 NATIONAL ELECTRICAL CODE SUMMARY FOR RESIDENTIAL WIRING

Receptacles that shall be GFCI protected in basements include all 125-volt through 250-volt receptacles supplied by single-phase branch circuits rated 150 volts or less to ground.

GFCI Protection for Indoor Damp and Wet Locations [210.8(A)(11)]

This section now applies to all dwelling indoor damp and wet locations. Receptacles that shall be GFCI protected in indoor damp and wet locations include all 125-volt through 250-volt receptacles supplied by single-phase branch circuits rated 150 volts or less to ground.

Island and Peninsular Countertops and Work Surfaces [210.52(C)(2)]

The minimum number of receptacle outlets required for island and peninsular countertops now depends on the square foot area of the countertop or work surface. At least one receptacle outlet shall be provided for the first 9 ft^2 (0.84 m^2), or fraction thereof, of the countertop or work surface. A receptacle outlet shall be provided for every additional 18 ft^2 (1.7 m^2), or fraction thereof, of the countertop or work surface. In a peninsular, at least one receptacle outlet shall be located within 2 ft (600 mm) of the outer end of a peninsular countertop or work surface. Additional required receptacle outlets can be located as determined by the installer, designer, or building owner.

Lighting Outlet Controlled by a Listed Wall-Mounted Control Device [210.70]

The term wall switch-controlled lighting outlet was replaced by the term lighting outlet controlled by a listed wall-mounted control device.

Point of Control [210.70(C)]

For attics and underfloor spaces, utility rooms, and basements, this section says at least one lighting outlet containing a switch or lighting controlled by a wall switch or listed wall-mounted control device shall be installed where these spaces are used for storage or contain equipment requiring servicing. This section now clarifies the term "point of control." A point of control shall be each entry that permits access to the attic and underfloor space, utility room, or basement. The last sentence was also reworded for clarity. Where a lighting outlet is installed for equipment requiring service, the lighting outlet shall be installed at or near the equipment.

Surge Protection [230.67]

As stated in 230.67(A), all services supplying dwelling units shall be provided with a surge-protective device (SPD). Other provisions in this section pertain to location, type, and service equipment that is replaced.

Maximum Number of Disconnects [230.71]

Each service shall have only one disconnecting means unless the requirements of 230.71(B) are met. Installing a main lug only (MLO) panelboard with not more than six circuit breakers used as disconnect switches will now be a violation. As stated in 230.71 (B)(2), separate enclosures with a main service disconnecting means in each enclosure shall be permitted.

Emergency Disconnects [230.85]

For one- and two-family dwelling units, all service conductors shall terminate in disconnecting means having a short-circuit current rating equal to or greater than the available fault current installed in a readily accessible outdoor location. If more than one disconnect is provided, they shall be grouped. In previous editions, the service disconnect(s) could be located outside the building or inside nearest the point of entrance. The service disconnecting means for one- and two-family dwelling units is now required to be installed outdoors. The service disconnecting means is still required to be in a readily accessible outdoor location. See 230.85(1) through (3) for types of disconnects and the required makings.

PARTIAL 2020 NATIONAL ELECTRICAL CODE SUMMARY FOR RESIDENTIAL WIRING

Conductors for General Wiring [Article 310]

Article 310 was reorganized to increase the usability of the article. Table 310.15(B)(16) was changed to Table 310.16. This table, which covers ampacities of insulated conductors, is one of the most used tables in the *NEC*®.

Receptacle Orientation Under Sinks [406.5(G)(2)]

Receptacles shall not be installed in a face-up position in the area below a sink.

Tamper-Resistant Receptacles [406.12]

Required locations for tamper-resistant receptacles have been expanded to include attached and detached garages and accessory buildings to dwelling units, common areas of multifamily dwellings, common areas of guest rooms and guest suites of hotels and motels, and assisted living facilities.

GFCI Protection for Swimming Pool Motors [680.21(C)]

Outlets supplying all pool motors on branch circuits rated 150 volts or less to ground and 60 amperes or less, single- or 3-phase shall be provided with Class A ground-fault circuit-interrupter protection. There is an exception for listed low-voltage motors.

Pool Pump Motor Replacement [680.21 (D)]

Where a pool pump motor in 680.21(C) is replaced for maintenance or repair, the replacement pump motor shall be provided with ground-fault circuit interrupter protection.

🔧 FIELD TERMS VERSUS *NEC*® TERMS

• BX	Armored cable (*NEC*® 320)
• Romex	Nonmetallic sheathed cable (*NEC*® 334)
• Greenfield	Flexible metal conduit (*NEC*® 348)
• Thin wall	Electrical metallic tubing (*NEC*® 358)
• Smurf tube	Electrical nonmetallic tubing (*NEC*® 362)
• 1900 box	4-inch square box (*NEC*® 314)
• 4-S box	4-inch square box (*NEC*® 314)
• 5-S box	$4^{11}/_{16}$-inch square box (*NEC*® 314)
• 333 box	Device box (*NEC*® 314)
• EYS	Explosion proof seal off (*NEC*® 500)
• Neutral**	Grounded conductor (*NEC*® 200)**
• Ground wire	Equipment grounding conductor (*NEC*® 250.118)
• Ground wire	Grounding electrode conductor (*NEC*® 250.66)
• Hot, live	Energized, or energized conductor (*NEC*® 100)
• Condulet	Conduit body (*NEC*® 300.15)
• Flex	Flexible metal conduit, Type FMC (*NEC*® 348)
• Guts	Panelboard interior (*NEC*® 100)
• Heavy-wall	Rigid metal conduit, Type RMC (*NEC*® 344)
• Red Head	Anti-short bushing for AC cable (*NEC*® 320)
• Sealtigh	Liquidtight flexible metal conduit, Type LFMC ((*NEC*® 350)
• Wiremold	Surface metal raceway ((*NEC*® 386
• Kearny or Bug	Split-bolt connector ((*NEC*® 110)

**Some systems do not have a neutral and the grounded conductor may be a phase conductor. (See *NEC*® Article 100 neutral definition.)

Home run to panel.
  Number of arrows indicates number of circuits.
Full slashes indicate ungrounded "hot"
(or switch leg) circuit conductors. Half slash
indicates grounded (neutral) circuit conductor.
No slashes indicate one "hot" and one neutral.

Branch circuit.
  Full slashes indicate ungrounded "hot" (or switch
leg) conductors. Half slash indicates grounded
(neutral) circuit conductor. No slashes indicate one
"hot" and one neutral.

Wiring concealed in finished areas,
exposed in unfinished areas

Switch wiring

Cable or conduit turning up

Cable or conduit turning down

🔌 PLAN SYMBOLS

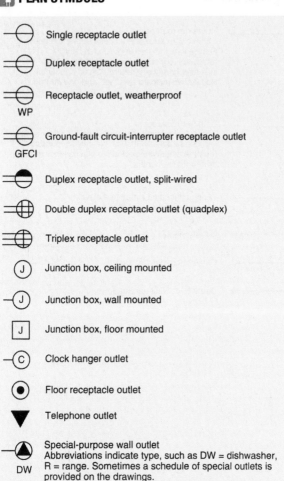

—⊖	Single receptacle outlet
=⊖	Duplex receptacle outlet
=⊖ WP	Receptacle outlet, weatherproof
=⊖ GFCI	Ground-fault circuit-interrupter receptacle outlet
=⊖	Duplex receptacle outlet, split-wired
=⊕	Double duplex receptacle outlet (quadplex)
=⊕	Triplex receptacle outlet
Ⓙ	Junction box, ceiling mounted
—Ⓙ	Junction box, wall mounted
☐J	Junction box, floor mounted
—Ⓒ	Clock hanger outlet
⊙	Floor receptacle outlet
▼	Telephone outlet
—▲ DW	Special-purpose wall outlet Abbreviations indicate type, such as DW = dishwasher, R = range. Sometimes a schedule of special outlets is provided on the drawings.

PLAN SYMBOLS

○	Incandescent, compact fluorescent, LED, or high-intensity discharge (H.I.D.) luminaire, surface mounted
—○	Incandescent, compact fluorescent, LED, or H.I.D. luminaire, wall mounted
Ⓡ	Incandescent, compact fluorescent, LED, or H.I.D. luminaire, recessed
▭○▭	Fluorescent luminaire, surface mounted
▭○R▭	Fluorescent luminaire, recessed
⊢○⊣	Bare-lamp fluorescent strip luminaire, surface mounted
S	Single-pole switch
S_2	Double-pole switch
S_3	Three-way switch
S_4	Four-way switch
S_{DS}	Dimmer switch
S_G	Glow snap switch, glows in off position
S_P	Switch with pilot light
S_T	Timer switch
S_{WP}	Single-pole switch, weatherproof

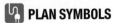

	Panelboard, recessed
	Panelboard, surface mounted
(M)	Motor
▼	Telephone outlet
	Disconnect switch
	Ground
	Circuit breaker
Ø	Phase

⚡ ELECTRICAL SYMBOLS

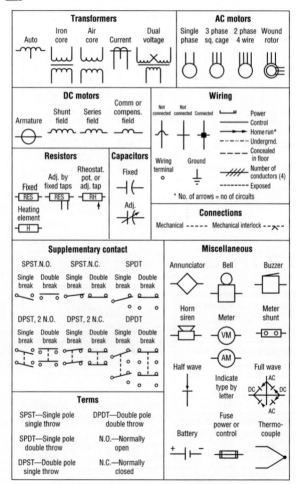

Transformers

Auto — Iron core — Air core — Current — Dual voltage

AC motors

Single phase — 3 phase sq. cage — 2 phase 4 wire — Wound rotor

DC motors

Armature — Shunt field — Series field — Comm or compens. field

Wiring

Not connected — Not connected — Connected

Power
Control
Home run*
Undergrnd.
Concealed in floor
Number of conductors (4)
Exposed

Wiring terminal — Ground

* No. of arrows = no of circuits

Resistors

Fixed — Adj. by fixed taps — Rheostat. pot. or adj. tap — RES — RES — RH

Heating element — H

Capacitors

Fixed
Adj.

Connections

Mechanical ----- Mechanical interlock

Supplementary contact

SPST.N.O.
Single break — Double break

SPST.N.C.
Single break — Double break

SPDT
Single break — Double break

DPST, 2 N.O.
Single break — Double break

DPST, 2 N.C.
Single break — Double break

DPDT
Single break — Double break

Miscellaneous

Annunciator — Bell — Buzzer

Horn siren — Meter — Meter shunt

Half wave — Meter (VM) — (AM) Indicate type by letter

Full wave

Fuse power or control — Thermo-couple

Battery

Terms

SPST—Single pole single throw

SPDT—Single pole double throw

DPST—Double pole single throw

DPDT—Double pole double throw

N.O.—Normally open

N.C.—Normally closed

⚡ ELECTRICAL SYMBOLS

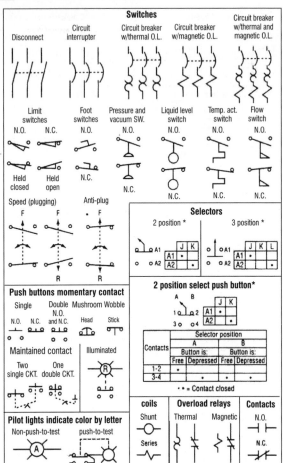

Switches

| Disconnect | Circuit interrupter | Circuit breaker w/thermal O.L. | Circuit breaker w/magnetic O.L. | Circuit breaker w/thermal and magnetic O.L. |

| Limit switches | Foot switches | Pressure and vacuum SW. | Liquid level switch | Temp. act. switch | Flow switch |

N.O. N.C.   N.O.   N.O.   N.O.   N.O.   N.O.

Held closed   Held open   N.C.   N.C.   N.C.   N.C.   N.C.

Speed (plugging)   Anti-plug

F   F   • F

R   R

Selectors

2 position *

		J	K
	A1	•	
	A2		•

3 position *

		J	K	L
	A1	•		
	A2			•

Push buttons momentary contact

| Single | Double N.O. and N.C. | Mushroom Head | Wobble Stick |
| N.O. N.C. | | | |

Maintained contact

Two single CKT.   One double CKT.

Illuminated

2 position select push button*

		J	K
	A1	•	
	A2		•

Contacts	Selector position			
	A		B	
	Button is:		Button is:	
	Free	Depressed	Free	Depressed
1-2	•			
3-4		•	•	•

• = Contact closed

Pilot lights indicate color by letter

Non-push-to-test   push-to-test

(A)

coils	Overload relays		Contacts
Shunt	Thermal	Magnetic	N.O.
Series			N.C.

Note: N.O. = normally open; N.C. = normally closed

124

🔌 WIRING DIAGRAMS FOR NEMA CONFIGURATIONS

2-Pole, 2-Wire Non-Grounding 125V

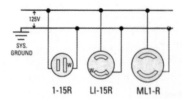

2-Pole, 2-Wire Non-Grounding 250V

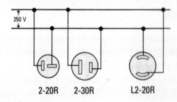

2-Pole, 3-Wire Grounding 125V

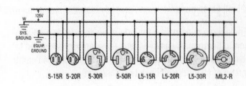

2-Pole, 3-Wire Grounding 250V

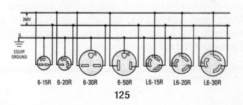

WIRING DIAGRAMS FOR NEMA CONFIGURATIONS

2-Pole, 3-Wire Grounding 277V AC

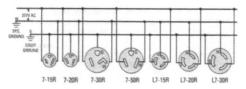

Courtesy of Cooper Wiring Devices.

 NEMA ENCLOSURE TYPES—NONHAZARDOUS LOCATIONS

The purpose of this document is to provide general information on the definitions of NEMA Enclosure Types to architects, engineers, installers, inspectors, and other interested parties. [For more detailed and complete information, NEMA Standards Publication 250-2018, "Enclosures for Electrical Equipment (1000 Volts Maximum)" should be consulted.]

In **non-hazardous locations,** the specific enclosure types, their applications, and the environmental conditions they are designed to protect against, when completely and properly installed, are as follows:

Type 1—Enclosures constructed for indoor use to provide a degree of protection to personnel against access to hazardous parts and to provide a degree of protection of the equipment inside the enclosure against ingress of solid foreign objects (falling dirt).

Type 2—Enclosures constructed for indoor use to provide a degree of protection to personnel against access to hazardous parts; to provide a degree of protection of the equipment inside the enclosure against ingress of solid foreign objects (falling dirt); and to provide a degree of protection with respect to harmful effects on the equipment due to the ingress of water (dripping and light splashing).

Reprinted by permission of the National Electrical manufacturers Association, Electrical Enclosure Types - Non Hazardous Location Environmental Rating Standards Comparison.

🔌 NEMA ENCLOSURE TYPES—NONHAZARDOUS LOCATIONS

Type 3—Enclosures constructed for either indoor or outdoor use to provide a degree of protection to personnel against access to hazardous parts; to provide a degree of protection of the equipment inside the enclosure against ingress of solid foreign objects (falling dirt and windblown dust); to provide a degree of protection with respect to harmful effects on the equipment due to the ingress of water (rain, sleet, snow); and against damage by the external formation of ice on the enclosure.

Type 3R—Enclosures constructed for either indoor or outdoor use to provide a degree of protection to personnel against access to hazardous parts; to provide a degree of protection of the equipment inside the enclosure against ingress of solid foreign objects (falling dirt); to provide a degree of protection with respect to harmful effects on the equipment due to the ingress of water (rain, sleet, snow); and against damage by the external formation of ice on the enclosure.

Type 3S—Enclosures constructed for either indoor or outdoor use to provide a degree of protection to personnel against access to hazardous parts; to provide a degree of protection of the equipment inside the enclosure against ingress of solid foreign objects (falling dirt and windblown dust); to provide a degree of protection with respect to harmful effects on the equipment due to the ingress of water (rain, sleet, snow); and for which the external mechanism(s) remain(s) operable when ice laden.

Type 3X—Enclosures constructed for either indoor or outdoor use to provide a degree of protection to personnel against access to hazardous parts; to provide a degree of protection of the equipment inside the enclosure against ingress of solid foreign objects (falling dirt and windblown dust); to provide a degree of protection with respect to harmful effects on the equipment due to the ingress of water (rain, sleet, snow); to provide an increased level of protection against corrosion and against damage by the external formation of ice on the enclosure.

Reprinted from *NEMA 250-2018* by permission of the National Electrical Manufacturers Association.

**NEMA ENCLOSURE TYPES—NONHAZARDOUS
LOCATIONS**

Type 3RX—Enclosures constructed for either indoor or outdoor use to provide a degree of protection to personnel against access to hazardous parts; to provide a degree of protection of the equipment inside the enclosure against ingress of solid foreign objects (falling dirt); to provide a degree of protection with respect to harmful effects on the equipment due to the ingress of water (rain, sleet, snow); to protect against damage by the external formation of ice on the enclosure and to provide an increased level of protection against corrosion.

Type 3SX—Enclosures constructed for either indoor or outdoor use to provide a degree of protection to personnel against access to hazardous parts; to provide a degree of protection of the equipment inside the enclosure against ingress of solid foreign objects (falling dirt and windblown dust); to provide a degree of protection with respect to harmful effects on the equipment due to the ingress of water (rain, sleet, snow); to provide an increased level of protection against corrosion; and for which the external mechanism(s) remain(s) operable when ice laden.

Type 4—Enclosures constructed for either indoor or outdoor use to provide a degree of protection to personnel against access to hazardous parts; to provide a degree of protection of the equipment inside the enclosure against ingress of solid foreign objects (failing dirt and windblown dust); to provide a degree of protection with respect to harmful effects on the equipment due to the ingress of water (rain, sleet, snow, splashing water, and hose-directed water); and against damage by the external formation of ice on the enclosure.

Type 4X—Enclosures constructed for either indoor or outdoor use to provide a degree of protection to personnel against access to hazardous parts; to provide a degree of protection of the equipment inside the enclosure against ingress of solid foreign objects (falling dirt and windblown dust); to provide a degree of protection with respect to harmful effects on the equipment due to the ingress of water (rain,

Reprinted from *NEMA 250-2018* by permission of the National Electrical Manufacturers Association.

sleet, snow, splashing water, and hose-directed water); to provide an increased level of protection against corrosion; and to protect against damage by the external formation of ice on the enclosure.

Type 5—Enclosures constructed for indoor use to provide a degree of protection to personnel against access to hazardous parts; to provide a degree of protection of the equipment inside the enclosure against ingress of solid foreign objects (falling dirt and settling airborne dust, lint, fibers, and flyings); and to provide a degree of protection with respect to harmful effects on the equipment due to the ingress of water (dripping and light splashing).

Type 6—Enclosures constructed for either indoor or outdoor use to provide a degree of protection to personnel against access to hazardous parts; to provide a degree of protection of the equipment inside the enclosure against ingress of solid foreign objects (falling dirt); to provide a degree of protection with respect to harmful effects on the equipment due to the ingress of water (hose-directed water and the entry of water during occasional temporary submersion at a limited depth); and to protect against damage by the external formation of ice on the enclosure.

Type 6P—Enclosures constructed for either indoor or outdoor use to provide a degree of protection to personnel against access to hazardous parts; to provide a degree of protection of the equipment inside the enclosure against ingress of solid foreign objects (falling dirt); to provide a degree of protection with respect to harmful effects on the equipment due to the ingress of water (hose-directed water and the entry of water during prolonged submersion at a limited depth); to provide an increased level of protection against corrosion; and to protect against damage by the external formation of ice on the enclosure.

Reprinted from *NEMA 250-2018* by permission of the National Electrical Manufacturers Association.

NEMA ENCLOSURE TYPES—NONHAZARDOUS LOCATIONS

Type 12—Enclosures constructed (without knockouts) for indoor use to provide a degree of protection to personnel against access to hazardous parts; to provide a degree of protection of the equipment inside the enclosure against ingress of solid foreign objects (falling dirt and circulating dust, lint, fibers, and flyings); to provide a degree of protection with respect to harmful effects on the equipment due to the ingress of water (dripping and light splashing); and to provide a degree of protection against light splashing and seepage of oil and noncorrosive coolants.

Type 12K—Enclosures constructed (with knockouts) for indoor use to provide a degree of protection to personnel against access to hazardous parts; to provide a degree of protection of the equipment inside the enclosure against ingress of solid foreign objects (falling dirt and circulating dust, lint, fibers, and flyings); to provide a degree of protection with respect to harmful effects on the equipment due to the ingress of water (dripping and light splashing); and to provide a degree of protection against light splashing and seepage of oil and noncorrosive coolants.

Type 13—Enclosures constructed for indoor use to provide a degree of protection to personnel against access to hazardous parts; to provide a degree of protection of the equipment inside the enclosure against ingress of solid foreign objects (falling dirt and circulating dust, lint, fibers, and flyings); to provide a degree of protection with respect to harmful effects on the equipment due to the ingress of water (dripping and light splashing); and to provide a degree of protection against the spraying, splashing, and seepage of oil and noncorrosive coolants.

Reprinted from *NEMA 250-2018* by permission of the National Electrical Manufacturers Association.

⚡ SINGLE-PHASE MOTORS

Split-Phase, Squirrel-Cage, Dual-Voltage Motor

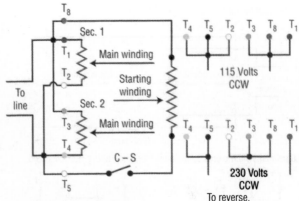

115 Volts
CCW

230 Volts
CCW

To reverse,
interchange 5 and 8

Classes of Single-Phase Motors

1. Split-phase
   A. Capacitor start
   B. Repulsion start
   C. Resistance start
   D. Split capacitor

2. Commutator
   A. Repulsion
   B. Series

Terminal Color Marking

| T_1 Blue • | T_3 Orange ○ | T_5 Black • |
| T_2 White | T_4 Yellow | T_8 Bed • |

Note: Split-phase motors are usually fractional horsepower. The majority of electric motors used in washing machines, refrigerators, etc. are of the split-phase type.

To change the speed of a split-phase motor, the number of poles must be changed.
1. Addition of running winding
2. Two starting windings and two running windings
3. Consequent pole connections

Split-Phase, Squirrel-Cage Motor

A. Resistance Start:

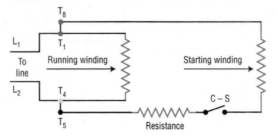

Centrifugal switch o (cs) opens after reaching 75% of normal speed.

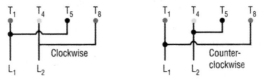

B. Capacitor Start:

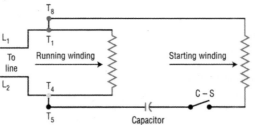

Note:
1. A resistance start motor has a resistance connected in series with the starting winding.
2. The capacitor start motor is employed where a high starting torque is required

132

SINGLE-PHASE USING STANDARD THREE-PHASE STARTER

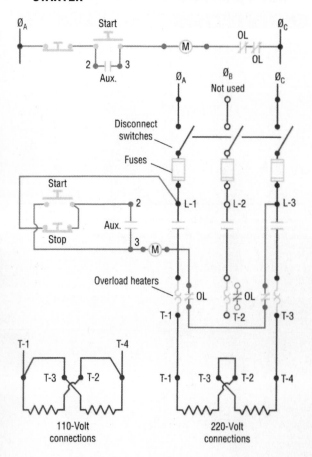

110-Volt connections

220-Volt connections

(M) = Motor starter coil

133

FULL-LOAD CURRENT FOR SINGLE-PHASE ALTERNATING CURRENT MOTORS (AMPERES)

HP	115 V	200 V	208 V	230 V
⅙	4.4	2.5	2.4	2.2
¼	5.8	3.3	3.2	2.9
⅓	7.2	4.1	4.0	3.6
½	9.8	5.6	5.4	4.9
¾	13.8	7.9	7.6	6.9
1½	16	9.2	8.8	8.0
1½	20	11.5	11	10
2	24	13.8	13.2	12
3	34	19.6	18.7	17
5	56	32.2	30.8	28
7½	80	46	44	40
10	100	57.5	55	50

The voltages listed are rated motor voltages. The currents listed shall be permitted for system voltage ranges of 110 to 120 and 220 to 240 volts.
Source: NFPA 70®, *National Electrical Code*®, NFPA, Quincy, MA, 2020, Table 430.248.

RUNNING OVERLOAD UNITS

Kind of Motor	Supply System	Number and Location of Overload Units Such as Trip Coils or Relays
1-phase ac or dc	2-wire, 1-phase ac or dc, ungrounded	1 in either conductor
1-phase ac or dc	2-wire, 1-phase ac or dc, one conductor grounded	1 in ungrounded conductor
1-phase ac or dc	3-wire, 1-phase ac ordc, grounded neutral conductor	1 in either ungrounded conductor
1-phase ac	Any 3-phase	1 in ungrounded conductor
2-phase ac	3-wire, 2-phase ac, ungrounded	2, one in each phase
2-phase ac	3-wire, 2-phase ac, one conductor grounded	2 in ungrounded conductors
2-phase ac	4-wire, 2-phase ac, grounded or ungrounded	2, onefor each phase in ungrounded conductors
2-phase ac	Grounded neutral or 5-wire, 2-phase ac, ungrounded	2, one for each phase in any ungrounded phase wire
3-phase ac	Any 3-phase	3, one in each phase*

*Exception: An overload unit in each phase shall not be required where overload protection is provided by other approved means.
Source: NFPA 70®, *National Electrical Code*®, NFPA, Quincy, MA, 2020, Table 430.37.

MOTOR BRANCH-CIRCUIT PROTECTIVE DEVICES: MAXIMUM RATING OR SETTING

Type of Motor	Percent of Full-Load Current			
	Nontime Delay Fuse[1]	Dual Element (Time-Delay) Fuse[1]	Instan-taneous Trip Breaker	Inverse Time Breaker[2]
Single-phase motors	300	175	800	250
AC polyphase motors other than wound rotor	300	175	800	250
Squirrel cage—other than Design B energy-efficient	300	175	800	250
Design B energy-efficient	300	175	1100	250
Synchronous[3]	300	175	800	250
Wound rotor	150	150	800	150
Direct current (constant voltage)	150	150	250	150

For certain exceptions to the values specified, see 430.54.
1. The values in the Nontime Delay Fuse column apply to Time-Delay Class CC fuses.
2. The values given in the last column also cover the ratings of nonadjustable inverse time types of circuit breakers that may be modified as in 430.52(C)(1), Exception No. 1 and No. 2.
3. Synchronous motors of the low-torque, low-speed type (usually 450 rpm or lower), such as are used to drive reciprocating compressors, pumps, and so forth, that start unloaded, do not require a fuse rating or circuit-breaker setting in excess of 200 % of full-load current.

Source: NFPA 70®, *National Electrical Code*®, NFPA, Quincy, MA, 2020, Table 430.52.
Note: Where the result of the calculation for the branch circuit protective device does not correspond with a standard size fuse or circuit breaker, see 430.52(C) Exception No. 1.
Note: Where the rating specified in Table 430.52, or the rating modified by Exception No. 1, is not sufficient for the starting current of the motor, see 430.52(C) Exception No. 2.

 MOTOR AND MOTOR CIRCUIT CONDUCTOR PROTECTION

Motors can have large starting currents three to five times or more than that of the actual motor current. In order for motors to start, the motor and motor circuit conductors are allowed to be protected by circuit breakers and fuses at values that are higher than the actual motor and conductor ampere ratings. These larger overcurrent devices do not provide overload protection and will only open upon short circuits or ground faults. Overload protection must be used to protect the motor based on the actual nameplate amperes of the motor. This protection is usually in the form of heating elements in manual or magnetic motor starters. Small motors, such as waste disposal motors, have a red overload reset button built into the motor.

General Motor Rules

- Use full-load current from tables instead of nameplate.
- Branch circuit conductors: Use 125% of full-load current to find conductor size.
- Branch circuit OCP size: Use percentages given in tables for full-load current. (*Ugly's* page 134)

Circuit Breaker Used for Overload Protection

Instead of installing a motor controller (or starter) with motor overload protection, it is permissible to install an overcurrent device (fuse or breaker) that provides overload protection for the motor and also provides protection for the conductors. Where an inverse time circuit breaker is also used for overload protection, it shall conform to the appropriate provisions of Article 430 governing overload protection [430.83(A)(2)],

SINGLE-PHASE TRANSFORMER CONNECTIONS

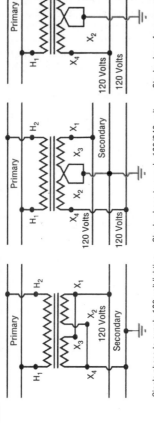

Single phase to supply 120-volt lighting load. Often used for single customer.

Single phase to supply 120/240-volt, 3-wire lighting and power load. Used in urban distribution circuits.

Single phase for power. Used for small industrial applications.

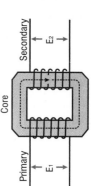

Single Ø transformer circuit

A transformer is a stationary induction device for transferring electrical energy from one circuit to another without change of frequency. A transformer consists of two coils or windings wound upon a magnetic core of soft iron laminations and insulated from one another

138

⚡ FULL-LOAD CURRENTS FOR SINGLE-PHASE TRANSFORMERS

kVA Rating	Single-Phase Transformer's Voltage				
	120	208	240	480	2400
1	8.33	4.81	4.17	2.08	.42
3	25.0	14.4	12.5	6.25	1.25
5	41.7	24.0	20.8	10.4	2.08
7.5	62.5	36.1	31.3	15.6	3.13
10	83.3	48.1	41.7	20.8	4.17
15	125.0	72.1	62.5	31.3	6.25
25	208.3	120.2	104.2	52.1	10.4
37.5	312.5	180.3	156.3	78.1	15.6
50	416.7	240.4	208.3	104.2	20.8
75	625.0	360.6	312.5	156.3	31.3
100	833.3	480.8	416.7	208.3	41.7
125	1041.7	601.0	520.8	260.4	52.1
167.5	1395.8	805.3	697.9	349.0	69.8
200	1666.7	961.5	833.3	416.7	83.3
250	2083.3	1201.9	1041.7	520.8	104.2
333	2775.0	1601.0	1387.5	693.8	138.8
500	4166.7	2403.8	2083.3	1041.7	208.3

$$I = \frac{kVA \times 1000}{E} \quad \text{or} \quad kVA = \frac{E \times I}{1000}$$

⚡ TRANSFORMER CALCULATIONS

To better understand the following formulas, review the rule of transposition in equations.

A multiplier may be removed from one side of an equation by making it a divisor on the other side; or a divisor may be removed from one side of an equation by making it a multiplier on the other side.

1. Voltage and Current: Primary (p) and Secondary (s)

Power (p) = Power (s)   or $E_p \times I_p = E_s \times I_s$

A. $E_p = \dfrac{E_s \times I_s}{I_p}$

B. $I_p = \dfrac{E_s \times I_s}{E_p}$

C. $\dfrac{E_p \times I_p}{E_s} = I_s$

D. $\dfrac{E_p \times I_p}{I_s} = E_s$

2. Voltage and Turns in Coil:

Voltage (p) x Turns (s) = Voltage (s) x Turns (p)

or

$E_p \times T_s = E_s \times T_p$

A. $E_p = \dfrac{E_s \times T_p}{T_s}$

B. $T_s = \dfrac{E_s \times T_p}{E_p}$

C. $\dfrac{E_p \times T_s}{E_s} = T_p$

D. $\dfrac{E_p \times T_s}{T_p} = E_s$

3. Amperes and Turns in Coil:

Amperes (p) x Turns (p) = Amperes (s) x Turns (s)

or

$I_p \times T_p = I_s \times T_s$

A. $I_p = \dfrac{I_s \times T_s}{T_p}$

B. $T_p = \dfrac{I_s \times T_s}{I_p}$

C. $\dfrac{I_p \times T_p}{I_s} = T_s$

D. $\dfrac{I_p \times T_p}{T_s} = I_s$

⚡ SIZING TRANSFORMERS

Single-Phase Transformers

Size a 480-volt primary to 240/120-volt secondary single-phase transformer for the following single-phase incandescent lighting load consisting of 48 recessed fixtures each rated 2 amperes, 120 volt. Each fixture has a 150-watt lamp. These fixtures can be evenly balanced on the transformer.

Find total volt-amperes using fixture ratings—*do not use lamp watt rating.*

$$2 \text{ amperes} \times 120 \text{ volts} = 240\text{-volt amperes}$$
$$240\text{-VA} \times 48 = 11520 \text{ VA}$$
$$11520 \text{ VA}/1000 = 11.52 \text{ KVA}$$

The single-phase transformer that meets or exceeds this value is **15 KVA.**

* 24 lighting fixtures would be connected line one to the common neutral and the other 24 lighting fixtures would be connected line two to the common neutral.

⚡ SINGLE-PHASE TRANSFORMER PRIMARY AND SECONDARY AMPERES

A 480/240-volt single-phase 50 KVA transformer ($Z = 2\%$) is to be installed. Calculate primary and secondary amperes, and short circuit amperes

Primary Amperes:

$$I_p = \frac{\text{KVA} \times 1000}{E_p} = \frac{50 \times 1000}{480} = \frac{50000}{480} = \textbf{104 amperes}$$

Secondary Amperes:

$$I_s = \frac{\text{KVA} \times 1000}{E_s} = \frac{50 \times 1000}{240} = \frac{50000}{240} = \textbf{208 amperes}$$

Short Circuit Amperes:

$$I_{sc} = \frac{I_s}{\%Z} = \frac{208}{.02} = \textbf{10400 amperes}$$

BUCK AND BOOST TRANSFORMER CONNECTIONS

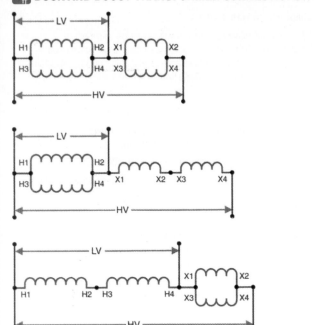

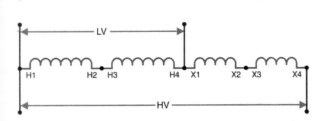

⚡ TYPICAL APPLIANCE POWER REQUIREMENTS

Load	Running Watts	Starting Watts	Load	Running Watts	Starting Watts
Lights	5-150	0	Washing machine	1200	2300
Security system	250	0	Fan, attic	350	400
Refrigerator	900	2300	Fan, bathroom	100	0
Freezer	900	2300	Fan, ceiling	300	200
Microwave	1450	0	Fan, fireplace	300	200
Furnace blower, ⅛ hp	300	500	Fan, kitchen exhaust	300	200
Furnace blower, ⅙ hP	500	900	Garage door opener, ¼ hp	550	1100
Furnace blower, ¼ hp	600	1100	Garage door opener, ⅓ hp	800	1700
Furnace blower, ⅓ hp	800	1400	Garage door opener, ½ hp	1000	2300
Furnace blower, ½ hp	1000	2350	Electric blanket	400	0
Pump, ⅓ hp	800	1400	Dehumidifier, portable	650	800
Pump, ½ hp	1050	2150	Vacuum cleaner	800-1100	0
Pump, ¾ hp	1400	2200	Coffeemaker	1750	0
Pump, 1 hp	2000	3000	Toaster	1050-1650	0
Electric range, 6 in. (152 mm) element	1500	0	Iron	1200	0 (152 mm) element
Electric range, 8 in. (203 mm) element	2100	0	Water heater	4500	0 (203 mm) element
Oven	6000	0	Hair dryer	300-1200	0
Air conditioner (window unit)	1000	2800	Television	100-500	0
Dishwasher (air-dry)	700	1400	Radio	50-200	0
Dishwasher (heat-dry)	1450	1400	Computer	600	0
Clothes dryer, gas	800	1800	Printer	220	0
Clothes dryer, electric	5750	1800	Copy machine	1500	0

⚡ GENERATOR SIZING

The generator must be capable of supplying the maximum connected load, in addition to the required starting current of the largest motor. The minimum generator ratings to serve the connected load are determined as follows (also see Generator Sizing Example):

1. *Determine the minimum generator wattage and ampere rating for the required toads.* The actual wattage of lighting and other loads are added, using nameplate information for appliances. When appliance nameplate information is not available, current and power requirements of the actual loads are measured or taken from page 143. The starting wattage required by the largest motor only (not all motors in the dwelling) is included in this calculation.

2. *Determine the minimum current rating of the generator* based on the power requirements of the connected load as determined in Step 1.

3. *First list the loads on a panelboard schedule and then balance them as closely as possible between phases.* This step is necessary because the generator must be sized to supply the largest required phase current of the connected load.

4. *Select a generator* with power and ampere ratings greater than the maximum connected load and with the maximum unbalanced load current.

Generator Sizing Example

This example determines the size of a generator needed to supply essential circuits only (not the entire dwelling unit load).

1. *Determine the minimum output power rating needed.* After calculating all the items that will be on this generator, the load will be 9250 watts.

2. *Select the generator.* A10-kilowatt (10000 watts) model would be sufficient to support the essential loads in this dwelling.

TELEPHONE SYSTEM PROTECTOR INSTALLATION RULES

The incoming telephone service cable is required to have a listed primary protector [805.90], which guards against high-voltage spikes caused by lightning. Sometimes this primary protector is installed by the telephone utility, but when installation is the responsibility of the electrical contractor, the steps that follow should be taken:

1. Install the type of listed primary protector supplied or specified by the local telephone utility. In many cases, this component comes built into the NIU/NID.

2. Connect a 14 AWG insulated solid or stranded grounding conductor (copper or other corrosion-resistant material) from the primary protector to the dwelling's grounding electrode system [800.100(A)], A telephone network interface installed indoors is often connected with a clamp to the metal water piping within 5 ft (1.5 m) of the point where this piping enters the building [800.100(B) and (C)].

3. Keep the grounding conductor under 20 ft (6.0 m) in length, and run it to the intersystem bonding termination in as straight a line as possible [800.100(A)(4) and (5) and 800.100(B)(1)]. In buildings or structures without an intersystem bonding termination, see 800.100(B)(2) or (3).

4. Follow the usual practice of running the grounding conductor exposed. If it is run in a metal raceway to protect it from damage, both ends of the raceway must be bonded to the conductor or to the same terminals/electrodes to which the grounding conductor is connected.

DEMARCATION POINT

The Demarcation Point Establishes Responsibility for Installing Telephone Wiring and Equipment. It Is Similar to the Service Point for Electric Light and Power

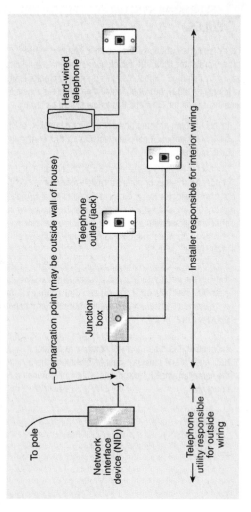

TELEPHONE CABLE COLOR CODE

Pair No.	Tip (+)	Ring (−)
1	White	Blue
2	White	Orange
3	White	Green
4	White	Brown
5	White	Slate
6	Red	Blue
7	Red	Orange
8	Red	Green
9	Red	Brown
10	Red	Slate
11	Black	Blue
12	Black	Orange
13	Black	Green
14	Black	Brown
15	Black	Slate
16	Yellow	Blue
17	Yellow	Orange
18	Yellow	Green
19	Yellow	Brown
20	Yellow	Slate
21	Violet	Blue
22	Violet	Orange
23	Violet	Green
24	Violet	Brown
25	Violet	Slate

TOP

PAIR CODE			SIDE #1		SIDE #2
Pair 1	Tip	26	White/Blue		White/Blue
	Ring	1	Blue/White		Blue/White
Pair 2	Tip	27	White/Orange		White/Orange
	Ring	2	Orange/White		Orange/White
Pair 3	Tip	28	White/Green		White/Green
	Ring	3	Green/White		Green/White
Pair 4	Tip	29	White/Brown		White/Brown
	Ring	4	Brown/White		Brown/White
Pair 5	Tip	30	White/Slate		White/Slate
	Ring	5	Slate/White		Slate/White
Pair 6	Tip	31	Red/Blue		Red/Blue
	Ring	6	Blue/Red		Blue/Red
Pair 7	Tip	32	Red/Orange		Red/Orange
	Ring	7	Orange/Red		Orange/Red
Pair 8	Tip	33	Red/Green		Red/Green
	Ring	8	Green/Red		Green/Red
Pair 9	Tip	34	Red/Brown		Red/Brown
	Ring	9	Brown/Red		Brown/Red
Pair 10	Tip	35	Red/Slate		Red/Slate
	Ring	10	Slate/Red		Slate/Red
Pair 11	Tip	36	Black/Blue		Black/Blue
	Ring	11	Blue/Black		Blue/Black
Pair 12	Tip	37	Black/Orange		Black/Orange
	Ring	12	Orange/Black		Orange/Black
Pair 13	Tip	38	Black/Green		Black/Green
	Ring	13	Green/Black		Green/Black
Pair 14	Tip	39	Black/Brown		Black/Brown
	Ring	14	Brown/Black		Brown/Black
Pair 15	Tip	40	Black/Slate		Black/Slate
	Ring	15	Slate/Black		Slate/Black
Pair 16	Tip	41	Yellow/Blue		Yellow/Blue
	Ring	16	Blue/Yellow		Blue/Yellow
Pair 17	Tip	42	Yellow/Orange		Yellow/Orange
	Ring	17	Orange/Yellow		Orange/Yellow
Pair 18	Tip	43	Yellow/Green		Yellow/Green
	Ring	18	Green/Yellow		Green/Yellow
Pair 19	Tip	44	Yellow/Brown		Yellow/Brown
	Ring	19	Brown/Yellow		Brown/Yellow
Pair 20	Tip	45	Yellow/Slate		Yellow/Slate
	Ring	20	Slate/Yellow		Slate/Yellow
Pair 21	Tip	46	Violet/Blue		Violet/Blue
	Ring	21	Blue/Violet		Blue/Violet
Pair 22	Tip	47	Violet/Orange		Violet/Orange
	Ring	22	Orange/Violet		Orange/Violet
Pair 23	Tip	48	Violet/Green		Violet/Green
	Ring	23	Green/Violet		Green/Violet
Pair 24	Tip	49	Violet/Brown		Violet/Brown
	Ring	24	Brown/Violet		Brown/Violet
Pair 25	Tip	50	Violet/Slate		Violet/Slate
	Ring	25	Slate/Violet		Slate/Violet

▣ TELEPHONE WIRING

Inside Wiring

Wire	Pair	Color Code
2-Pair	1	Green, Red
2-Pair	2	Black, Yellow
3-Pair	1	White/Blue, Blue/White
3-Pair	2	White/Orange, Orange/White
3-Pair	3	White/Green, Green/White

Inside Wire Terminations

Wire	Color	Function
2-Pair	Green,	Tip
2-Pair	Red	Ring
3-Pair	Black,	Optional
3-Pair	Yellow	Optional
3-Pair	White/Blue,	Tip
3-Pair	Blue/White	Ring
3-Pair	White/Orange,	Optional
3-Pair	Orange/White	Optional

Supports

Type	Spacing	From Corner
Wire-Clamp	16 in.	2 in.
Bride Ring	4 in.	6 in.
Drive Ring	4 in.	6 in.
Staple	7 ½ in.	2 in.

DATA CABLING COLOR CODE

TIA-606-B

Blue—Horizontal voice cables
Brown—Inter-building backbone
Gray—Second-level backbone
Green—Network connections
Orange—Demarcation point
Purple—First-level backbone
Red—Key systems (telephone)
Silver or White—Horizontal data cables
Yellow—Auxiliary, maintenance, security

Category 5 Color Code

Pair 1—White/blue, blue
Pair 2—White/orange, orange
Pair 3—White/green, green
Pair 4—White/brown, brown
White/blue = White insulation with blue stripe

▣ TYPICAL RJ45 CAT. 5 JACK

Pin/Pair Assignments

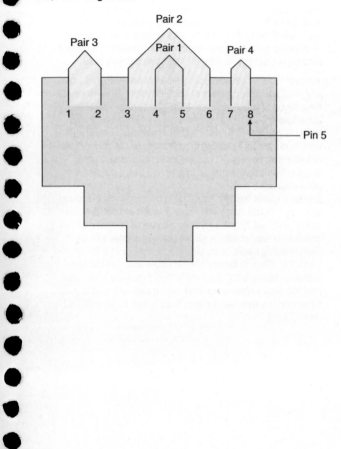

🔥 SMOKE DETECTOR PLACEMENT GUIDELINES

- *Avoid dead-air spaces.* Smoke detectors should be placed to avoid "dead-air spaces," which are not easily penetrated by smoke and heat. Units in finished rooms should not be placed close to locations where a wall meets the ceiling, because air does not circulate well in such locations. For the same reason, smoke detectors in unfinished basements should be mounted on, rather than between, joists.

- *Minimize nuisance alarms.* Cooking vapors or bathroom moisture can cause unwanted operation of smoke detectors. To minimize these so-called nuisance alarms, NFPA 72 requires smoke alarms and smoke detectors to be located not closer than 3 ft (910 mm) horizontal path from a door to a bathroom containing a shower or tub unless listed for installation in close proximity to such locations. Smoke alarms and smoke detectors shall not be installed within an area of exclusion determined by a 10 ft (3.0 m) radial distance along a horizontal flow path from a stationary or fixed cooking appliance, unless listed for installation in close proximity to cooking appliances. Smoke alarms and smoke detectors installed between 10 ft (3.0 m) and 20 ft (6.1 m) along a horizontal flow path from a stationary or fixed cooking appliance shall be equipped with an alarm-silencing means or use photoelectric detection [NFPA 72, 29.11.3.4], Effective January 1,2022, smoke alarms and smoke detectors installed 6 ft (1.8 m) and 20 ft (6.1 m) along a horizontal flow path from a stationary or fixed cooking appliance shall be listed for resistance to common nuisance sources from cooking [NFPA 72, 29.11.3.4(6)].

🔌 TYPICAL DOOR CHIME

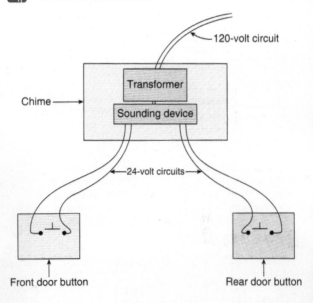

153

THERMOSTAT CIRCUIT

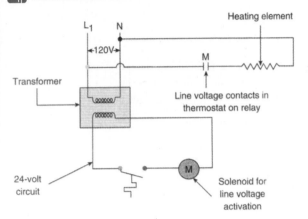

HVAC THERMOSTAT

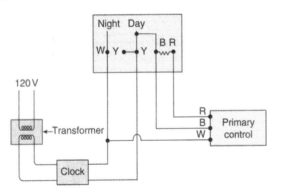

THREAD DIMENSIONS AND TAP DRILL SIZES

	Coarse Thread Series				Fine Thread Series		
Minimal Size	Threads per in.	Tap Drill	Clearance Drill	Nominal Size	Threads per in.	Tap Drill	Clearance Drill
1/64 in.	48	47	36	0	80	3/64 in.	51
1/16 in.	40	38	29	1	72	53	47
6	32	36	25	2	64	50	42
8	32	29	16	3	56	45	36
10	24	25		4	48	42	31
12	24	16	11/64 in.	1/8 in.	44	37	29
1/4 in.	20	7	7/32 in.	6	40	33	25
5/16 in.	18	F	17/64 in.	8	36	29	16
3/8 in.	16	5/16 in.	25/64 in.	10	32	21	
7/16 in.	14	U	29/64 in.	12	28	14	7/32 in.
1/2 in.	13	27/64 in.	33/64 in.	1/4 in.	28	3	17/64 in.
9/16 in.	12	31/64 in.	37/64 in.	5/16 in.	24	1	27/64 in.
5/8 in.	11	17/32 in.	41/64 in.	3/8 in.	24	Q	25/64 in.
3/4 in.	10	21/32 in.	49/64 in.	7/16 in.	20	25/64 in.	29/64 in.
7/8 in.	9	49/64 in.	57/64 in.	1/2 in.	20	29/64 in.	33/64 in.
1 in.	8	7/8 in.	1 1/64 in.	9/16 in.	18	33/64 in.	37/64 in.
1 1/8 in.	7	63/64 in.	1 11/64 in.	5/8 in.	18	37/64 in.	41/64 in.
1 1/4 in.	7	1 7/64 in.	1 17/64 in.	3/4 in.	16	11/16 in.	49/64 in.
1 1/2 in.	6	1 11/32 in.	1 25/64 in.	7/8 in.	14	13/16 in.	57/64 in.
2 in.	4 1/2	1 25/32 in.	2 1/32 in.	1	14	15/16 in.	1 1/64 in.

Hole Saw Chart*

Trade Size	Rigid Conduit	E.M.T. Conduit	Green-Field	L.T. Flex	Trade Size	Rigid Conduit	E.M.T. Conduit	Green-Field
1/2 in.	7/8 in.	3/4 in.	1 in.	1 1/8 in.	2 1/2 in.	3 in.	2 7/8 in.	2 7/8 in.
3/4 in.	1 1/8 in.	1 in.	1 1/8 in.	1 1/4 in.	3 in.	3 5/8 in.	3 1/2 in.	3 3/8 in.
1 in.	1 3/8 in.	1 1/4 in.	1 3/8 in.	1 1/4 in.	3 1/2 in.	4 1/8 in.	4 in.	4 in.
1 1/4 in.	1 3/4 in.	1 5/8 in.	1 3/4 in.	1 1/2 in.	4 in.	4 1/2 in.	4 1/2 in.	4 1/2 in.
1 1/2 in.	2 in.	1 7/8 in.	2 in.	2 in.	5 in.	5 5/8 in.		
2 in.	2 1/2 in.	2 3/8 in.	2 1/2 in.	2 1/4 in.	6 in.	6 5/8 in.		

Note: For an oil-type push button station, use size 1 1/32 in. knock-out punch.
* For connectors (male connectors and adapters), use Rigid Table

TIGHTENING TORQUE

In accordance with *NEC®* 110.14(D), tightening torque values for terminal connections shall be as indicated on equipment or in installation instructions provided by the manufacturer. An approved method shall be used to achieve the indicated torque value.

Examples of approved means of achieving the indicated torque values include torque tools or devices such as shear bolts or breakaway-style devices with visual indicators that demonstrate that the proper torque has been applied [110.14(D) Informational Note No. 1].

The equipment manufacturer can be contacted if numeric torque values are not indicated on the equipment or if the installation instructions are not available. *NEC®* Informative Annex I of UL Standard 486A-486B, Standard for Safety-Wire Connectors, provides torque values in the absence of manufacturer's recommendations [110.14(D) Informational Note No. 2],

Additional information for torqueing threaded connections and terminations can be found in Section 8.11 of NFPA 70B-2019, Recommended Practice for Electrical Equipment Maintenance [110.14(D) Informational Note No. 3].

 **DRILLING SPEED (RPM)**

Material	Bit Size (Inches)	Speed Range
Steel	⅜	1000-1500
Steel	¼	1500-2000
Steel	⅛	3000-4000
Cast Iron	⅜	1500-2000
Cast Iron	¼	2000-2500
Cast Iron	⅛	3500-4500
Brass	⅜	3000-3500
Brass	¼	4500-5000
Brass	⅛	6000-6500
Wood	Over 1	700-2000
Wood	¾	2200-3000
Wood	½	2700-3400
Wood	¼	3200-3800
Plastic	⅜	1500-2000
Plastic	¼	3000-3500
Plastic	⅛	5000-6000

 **SHEET METAL SCREWS**

Screw Size	Hole Size	Metal Gauges
4	$^{3}/_{32}$ in.	28–20
6	$^{7}/_{64}$ in.	28–20
8	$^{1}/_{8}$ in.	26–18
10	$^{9}/_{64}$ in.	26–18
12	$^{5}/_{32}$ in.	24–18
14	$^{3}/_{16}$ in.	24–18
$^{5}/_{16}$	$^{1}/_{4}$ in.	22–16

WOOD SCREWS

Screw Size	Length (Inches)	Pilot Hole (Hard Wood)	Pilot Hole (Soft Wood)	Shank Hole
2	¼-½	#65	#54	#43
4	⅜-¾	#55	#31	#32
6	⅜-1½	#52	#44	#27
8	½-2	#48	#35	#18
10	⅝-2½	#43	#31	#10
12	⅞-3½	#38	#25	#2
14	1-4½	#32	#14	#D
16	1¼-5½	#29	#10	#I
18	1½-6	#26	#6	#M
20	1¾-6	#19	#3	#P
24	3½-6	#15	#D	#V

STRENGTH OF NYLON ROPE

Diameter (Inches)	Diameter (mm)	Safe Load (Pounds)	Breaking Strength (Pounds)
3/16	5	72	880
¼	6	120	1460
5/16	8	190	2250
3/9	10	270	3200
7/16	11	360	4300
½	12	470	5700
9/16	14	600	7200
⅝	16	740	8900
¾	18	1050	12700
⅞	22	1450	17000
1	24	1850	22000
1¼	30	2900	35000
1½	36	4000	48000

Note: All figures approximate.

ALLOWABLE LOADS—PLASTIC ANCHORS: TENSION LOAD IN POUNDS

Anchor Size	Drill Hole	Depth	In Brick	In Block	In Concrete
#6 or 8 x $^3/_4$ in.	$^3/_{16}$ in.	$^3/_4$ in.	25	45	46
#8 or 10 x $^7/_8$ in.	$^3/_{16}$ in.	$^7/_8$ in.	40	73	75
#10 x 1 in.	$^3/_{16}$ in.	1 in.	50	78	81
#12 x 1 in.	$^1/_4$ in.	1 in.	70	88	88
#14 x 1½ in.	$^5/_{16}$ in.	1½ in.	200	180	180
#16 x 1½ in.	$^5/_{16}$ in.	1½ in.	220	210	210

Note: All figures approximate.

ALLOWED LOADS—CONCRETE SCREWS: TENSION LOAD IN POUNDS

Anchor Size	Drill Hole	Depth	In Brick	In Block	In Concrete
$^3/_{16}$ in.	$^5/_{32}$ in.	1 in.	155	160	165
$^3/_{16}$ in.	$^5/_{32}$ in.	1¼ in.	268	190	218
$^3/_{16}$ in.	$^5/_{32}$ in.	1½ in.	333	303	330
¼ in.	$^3/_{16}$ in.	1 in.	90	95	220
¼ in.	$^3/_{16}$ in.	1¼ in.	150	160	265
¼ in.	$^3/_{16}$ in.	1½ in.	162	179	440

Note: All figures approximate.

ALLOWABLE LOADS—EXPANSION SHIELDS: TENSION LOAD IN POUNDS

Anchor Size	Bolt Size	Drill Hole	Depth	In Concrete
$1/4$ in.	$1/4$ in.	$1/2$ in.	$1\,3/8$ in.	450
$1/4$ in.	$1/4$ in.	$1/2$ in.	$1\,1/2$ in.	470
$1/4$ in.	$1/4$ in.	$1/2$ in.	$1\,5/8$ in.	480
$5/16$ in.	$5/16$ in.	$5/8$ in.	$1\,5/8$ in.	540
$5/16$ in.	$5/16$ in.	$5/8$ in.	$1\,7/8$ in.	560
$5/16$ in.	$5/16$ in.	$3/4$ in.	$2\,1/8$ in.	650
$3/8$ in.	$3/8$ in.	$3/4$ in.	2 in.	800
$3/8$ in.	$3/8$ in.	$3/4$ in.	$2\,1/8$ in.	900
$1/2$ in.	$1/2$ in.	$7/8$ in.	$2\,1/2$ in.	1600
$1/2$ in.	$1/2$ in.	$7/8$ in.	$2\,5/8$ in.	1700

Note: All figures approximate.

⚡ OHM'S LAW

Ohm's Law is the relationship between voltage (E), current (I), and resistance (R). The rate of the current flow is equal to electromotive force divided by resistance.

i = **Intensity of Current = Amperes**
E = **Electromotive Force = Volts**
R = **Resistance = Ohms**
P = **Power = Watts**
The three basic Ohm's Law formulas are:

$$I = \frac{E}{R} \qquad\qquad R = \frac{E}{I} \qquad\qquad E = I \times R$$

Below is a chart containing the formulas related to Ohm's Law. To use the chart, start in the center circle and select the value you need to find, I (amps), R (ohms), E (volts), or P (watts). Then select the formula containing the values you know from the corresponding chart quadrant.

Example: An electrical appliance is rated at 1200 watts and is connected to 120 volts. How much current will it draw

Amperes $= \dfrac{\text{Watts}}{\text{Volts}}$ $\qquad I = \dfrac{P}{E}$ $\qquad I = \dfrac{1200}{120} = 10$ Amps

What is the resistance of the same appliance?

Ohms $= \dfrac{\text{Volts}}{\text{Amperes}}$ $\qquad R = \dfrac{E}{I}$ $\qquad R = \dfrac{120}{10} = 12$ Ohms

⚡ OHM'S LAW

In the preceding example, we know the following values:

I = amps = 10 Amps R = ohms = 12 Ohms
E = volts = 120 Volts P = watts = 1200 Watts

We can now see how the 12 formulas in the Ohm's Law chart can be applied.

$\text{Amps} = \sqrt{\dfrac{\text{Watts}}{\text{Ohms}}}$ $I = \sqrt{\dfrac{P}{R}} = \sqrt{\dfrac{1200}{12}} = \sqrt{100} = 10$ Amps

$\text{Amps} = \dfrac{\text{Watts}}{\text{Volts}}$ $I = \dfrac{P}{E} = \dfrac{1200}{120} = 10$ Amps

$\text{Amps} = \dfrac{\text{Volts}}{\text{Ohms}}$ $I = \dfrac{E}{R} = \dfrac{120}{12} = 10$ Amps

---

$\text{Watts} = \dfrac{\text{Volts}^2}{\text{Ohms}}$ $P = \dfrac{E^2}{R} = \dfrac{120^2}{12} = \dfrac{14400}{12} = 1200$ Watts

$\text{Watts} = \text{Volts} \times \text{Amps}$ $P = E \times I = 120 \times 10 = 1200$ Watts

$\text{Watts} = \text{Amps}^2 \times \text{Ohms}$ $P = I^2 \times R = 10^2 \times 12 = 1200$ Watts

---

$\text{Volts} = \sqrt{\text{Watts} \times \text{Ohms}}$ $E = \sqrt{P \times R} = \sqrt{1200 \times 12} = \sqrt{14400} = 120$ Volts

$\text{Volts} = \text{Amps} \times \text{Ohms}$ $E = I \times R = 10 \times 12 = 120$ Volts

$\text{Volts} = \dfrac{\text{Watts}}{\text{Amps}}$ $E = \dfrac{P}{I} = \dfrac{1200}{10} = 120$ Volts

---

$\text{Ohms} = \dfrac{\text{Volts}^2}{\text{Watts}}$ $R = \dfrac{E^2}{P} = \dfrac{120^2}{1200} = \dfrac{14400}{1200} = 12$ Ohms

$\text{Ohms} = \dfrac{\text{Watts}}{\text{Amps}^2}$ $R = \dfrac{P}{I^2} = \dfrac{1200}{10^2} = 12$ Ohms

$\text{Ohms} = \dfrac{\text{Volts}}{\text{Amps}}$ $R = \dfrac{E}{I} = \dfrac{120}{10} = 12$ Ohms

⚡ SERIES CIRCUITS

A series circuit is a circuit that has only one path through which the electrons may flow.

Rule 1: The total current In a series circuit is equal to the current in any other part of the circuit.

Total Current $I_T = I_1 = I_2 = I_3$, etc.

Rule 2: The total voltage in a series circuit is equal to the sum of the voltages across all parts of the circuit.

Total Voltage $E_T = E_1 + E_2 + E_3$, etc.

Rule 3: The total resistance of a series circuit is equal to the sum of the resistances of all the parts of the circuit.

Total Resistance $R_T = R_1 + R_2 + R_3$, etc.

Formulas from Ohm's Law

$$\text{Amperes} = \frac{\text{Volts}}{\text{Resistance}} \qquad \text{or} \qquad I = \frac{E}{R}$$

$$\text{Resistance} = \frac{\text{Volts}}{\text{Amperes}} \qquad \text{or} \qquad R = \frac{E}{I}$$

$$\text{Volts} = \text{Amperes} \times \text{Resistance} \qquad \text{or} \qquad E = I \times R$$

Example 1: Find the total voltage, total current, and total resistance of the following series circuit.

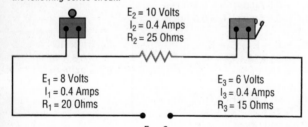

$E_2 = 10$ Volts
$I_2 = 0.4$ Amps
$R_2 = 25$ Ohms

$E_1 = 8$ Volts
$I_1 = 0.4$ Amps
$R_1 = 20$ Ohms

$E_3 = 6$ Volts
$I_3 = 0.4$ Amps
$R_3 = 15$ Ohms

$E_T = ?$
$I_T = ?$
$R_T = ?$

163

SERIES CIRCUITS

$$E_T = E_1 + E_2 + E_3$$
$$= 8 + 10 + 6$$
$$E_T = 24 \text{ Volts}$$

$$I_T = I_1 = I_2 = I_3$$
$$= 0.4 = 0.4 = 0.4$$
$$I_T = 0.4 \text{ Amps}$$

$$R_T = R_1 + R_2 + R_3$$
$$= 20 + 25 + 15$$
$$R_T = 60 \text{ Ohms}$$

Example 2: Find E_T, E_1, E_3, I_T, I_1, I_2, I_4, R_T, R_2, and R_4.
Remember that the total current in a series circuit is equal to the current in any other part of the circuit.

$E_1 = ?$ $E_3 = ?$
$I_1 = ?$ $I_3 = 0.5 \text{ Amps}$
$R_1 = 72 \text{ Ohms}$ $R_3 = 48 \text{ Ohms}$

$E_2 = 12 \text{ Volts}$ $E_4 = 48 \text{ Volts}$
$I_2 = ?$ $I_4 = ?$
$R_2 = ?$ $R_4 = ?$

$E_T = ?$ $I_T = ?$ $R_T = ?$

$$I_T = I_1 = I_2 = I_3 = I_4$$
$$I_T = I_1 = I_2 = 0.5 = I_4$$
$$0.5 = 0.5 = 0.5 = 0.5 = 0.5$$
$$I_T = 0.5 \text{ Amps} \quad I_2 = 0.5 \text{ Amps}$$
$$I_1 = 0.5 \text{ Amps} \quad I_4 = 0.5 \text{ Amps}$$

$$E_1 = I_1 \times R_1$$
$$= 0.5 \times 72$$
$$E_1 = 36 \text{ Volts}$$

$$E_T = E_1 + E_2 + E_3 + E_4$$
$$= 36 + 12 + 24 + 48$$
$$E_T = 120 \text{ Volts}$$

$$E_3 = I_3 \times R_3$$
$$= 0.5 \times 48$$
$$E_3 = 24 \text{ Volts}$$

$$R_T = R_1 + R_2 + R_3 + R_4$$
$$= 72 + 24 + 48 + 96$$
$$R_T = 240 \text{ Ohms}$$

$$R_2 = \frac{E_2}{I_2} = \frac{12}{0.5}$$
$$R_2 = 24 \text{ Ohms}$$

$$R_4 = \frac{E_4}{I_4} = \frac{48}{0.5}$$
$$R_4 = 96 \text{ Ohms}$$

164

 **PARALLEL CIRCUITS**

A parallel circuit is a circuit that has more than one path through which the electrons may flow.

Rule 1: The total current in a parallel circuit is equal to the sum of the currents in all the branches of the circuit.

Total Current $I_T = I_1 + I_2 + I_3$, etc.

Rule 2: The total voltage across any branch in parallel is equal to the voltage across any other branch and is also equal to the total voltage.

Total Voltage $E_T = E_1 = E_2 = E_3$, etc.

Rule 3: The total resistance of a parallel circuit is found by applying Ohm's Law to the total values of the circuit.

$$\text{Total Resistance} = \frac{\text{Total Voltage}}{\text{Total Amperes}} \quad \text{or} \quad R_T = \frac{E_T}{I_T}$$

Example: Find the total current, total voltage, and total resistance of the following parallel circuit.

$E_1 = 120$ Volts	$E_2 = 120$ Volts	$E_3 = 120$ Volts
$I_1 = 2$ Amps	$I_2 = 1.5$ Amps	$I_3 = 1$ Amps
$R_1 = 60$ Ohms	$R_2 = 80$ Ohms	$R_3 = 120$ Ohms

$$
\begin{aligned}
I_T &= I_1 + I_2 + I_3 \\
&= 2 + 1.5 + 1 \\
I_T &= 4.5 \text{ Amps}
\end{aligned}
\qquad
\begin{aligned}
E_T &= E_1 = E_2 = E_3 \\
&= 120 = 120 = 120 \\
E_T &= 120 \text{ Volts}
\end{aligned}
$$

$$R_T = \frac{E_T}{I_T} = \frac{120 \text{ Volts}}{4.5 \text{ Amps}} = 26.66 \text{ Ohms Resistance}$$

Note: In a parallel circuit, the total resistance is always less than the resistance of any branch. If the branches of a parallel circuit have the same resistance, then each will draw the same current. If the branches of a parallel circuit have different resistances, then each will draw a different current. In either series or parallel circuits, the larger the resistance, the smaller the current drawn.

COMBINATION CIRCUITS

In combination circuits, we combine series circuits with parallel circuits. Combination circuits make it possible to obtain the different voltages of series circuits and the different currents of parallel circuits

Example: Parallel-Series Circuit
Solve for all missing values.

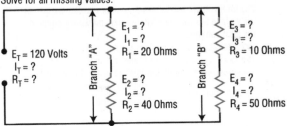

$E_T = 120$ Volts
$I_T = ?$
$R_T = ?$

Branch "A"

$E_1 = ?$
$I_1 = ?$
$R_1 = 20$ Ohms

$E_2 = ?$
$I_2 = ?$
$R_2 = 40$ Ohms

Branch "B"

$E_3 = ?$
$I_3 = ?$
$R_3 = 10$ Ohms

$E_4 = ?$
$I_4 = ?$
$R_4 = 50$ Ohms

To solve:
1. Find the total resistance of each branch. Both branches are simple series circuits, so:

$R_1 + R_2 = R_A$
$20 + 40 = 60$ Ohms total resistance of branch "A"

$R_3 + R_4 = R_B$
$10 + 50 = 60$ Ohms total resistance of branch "B"

2. Redraw the circuit, combining resistors ($R_1 + R_2$) and ($R_3 + R_4$) so that each branch will have only one resistor.

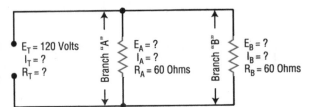

$E_T = 120$ Volts
$I_T = ?$
$R_T = ?$

Branch "A"

$E_A = ?$
$I_A = ?$
$R_A = 60$ Ohms

Branch "B"

$E_B = ?$
$I_B = ?$
$R_B = 60$ Ohms

COMBINATION CIRCUITS

Note: We now have a simple parallel circuit, so:

$$E_T = E_A = E_B$$
$$120 \text{ Volts} = 120 \text{ Volts} = 120 \text{ Volts}$$

We now have a parallel circuit with only two resistors, and they are of equal value. We have a choice of three different formulas that can be used to solve for the total resistance of the circuit.

(1) $\quad R_T = \dfrac{R_A \times R_B}{R_A + R_B} = \dfrac{60 \times 60}{60 + 60} = \dfrac{3600}{120} = 30 \text{ Ohms}$

(2) When the resistors of a parallel circuit are of equal value,

$\quad R_T = \dfrac{R}{N} = \dfrac{60}{2} = 30 \text{ Ohms} \qquad$ or

(3) $\quad \dfrac{1}{R_T} = \dfrac{1}{R_A} + \dfrac{1}{R_B} = \dfrac{1}{60} + \dfrac{1}{60} = \dfrac{2}{60} = \dfrac{1}{30}$

$\quad \dfrac{1}{R_T} \diagdown\!\!\!\!\diagup \dfrac{1}{30} \quad$ or $\quad 1 \times R_T = 1 \times 30 \quad$ or $\quad R_T = 30 \text{ Ohms}$

3. We know the values of E_T, R_T, E_A, R_A, E_B, R_B, R_1, R_2, R_3, and R_4. Next we will solve for I_T, I_A, I_B, I_1, I_2, I_3, and I_4.

$I_T = \dfrac{E_T}{R_T} \qquad$ or $\qquad \dfrac{120}{30} = 4 \qquad I_T = 4 \text{ Amps}$

$I_A = \dfrac{E_A}{R_A} \qquad$ or $\qquad \dfrac{120}{60} = 2 \qquad I_A = 2 \text{ Amps}$

$I_A = I_1 = I_2 \qquad$ or $2 = 2 = 2 \qquad \begin{aligned} I_1 &= 2 \text{ Amps} \\ I_2 &= 2 \text{ Amps} \end{aligned}$

$I_B = \dfrac{E_B}{R_B} = \qquad$ or $\qquad \dfrac{120}{60} = 2 \qquad I_B = 2 \text{ Amps}$

$I_B = I_3 = I_4 \qquad$ or $2 = 2 = 2 \qquad \begin{aligned} I_3 &= 2 \text{ Amps} \\ I_4 &= 2 \text{ Amps} \end{aligned}$

167

⚡ COMBINATION CIRCUITS

4. We know that resistors #1 and #2 of branch "A" are in series. We know too that resistors #3 and #4 of branch "B" are in series. We have determined that the total current of branch "A" is 2 amps, and the total current of branch "B" is 2 amps. By using the series formula, we can solve for the current of each branch.

Branch "A"	Branch "B"
$I_A = I_1 = I_2$	$I_B = I_3 = I_4$
$2 = 2 = 2$	$2 = 2 = 2$
$I_1 = 2$ Amps	$I_3 = 2$ Amps
$I_2 = 2$ Amps	$I_4 = 2$ Amps

5. We were given the resistance values of all resistors. $R_1 = 20$ Ohms, $R_2 = 40$ Ohms, $R_3 = 10$ Ohms, and $R_4 = 50$ Ohms. By using Ohm's Law, we can determine the voltage drop across each resistor.

$$E_1 = R_1 \times I_1$$
$$= 20 \times 2$$
$$E_1 = 40 \text{ Volts}$$

$$E_3 = R_3 \times I_3$$
$$= 10 \times 2$$
$$E_3 = 20 \text{ Volts}$$

$$E_2 = R_2 \times I_2$$
$$= 40 \times 2$$
$$E_2 = 80 \text{ Volts}$$

$$E_4 = R_4 \times I_4$$
$$= 50 \times 2$$
$$E_4 = 100 \text{ Volts}$$

ELECTRICAL FORMULAS FOR CALCULATING AMPERES, HORSEPOWER, KILOWATTS, AND kVA

To Find	Direct Current	Alternating Current		
		Single Phase	Two Phase/Four Wire	Three Phase
Amperes when "HP" is known	$\dfrac{HP \times 746}{E \times \%EFF}$	$\dfrac{HP \times 746}{E \times \%EFF \times PF}$	$\dfrac{HP \times 746}{E \times \%EFF \times PF \times 2}$	$\dfrac{HP \times 746}{E \times \%EFF \times PF \times 173}$
Amperes when "KW" is known	$\dfrac{KW \times 1000}{E}$	$\dfrac{KW \times 1000}{E \times PF}$	$\dfrac{KW \times 1000}{E \times PF \times 2}$	$\dfrac{KW \times 1000}{E \times PF \times 1.73}$
Amperes when "KVA" is known		$\dfrac{KVA \times 1000}{E}$	$\dfrac{KVA \times 1000}{E \times 2}$	$\dfrac{KVA \times 1000}{E \times 1.73}$
Kilowatts (Tru power)	$\dfrac{E \times I}{1000}$	$\dfrac{E \times I \times PF}{1000}$	$\dfrac{E \times I \times PF \times 2}{1000}$	$\dfrac{E \times I \times PF \times 1.73}{1000}$
Kilovolt-Amperes "kVA" (Apparent power)		$\dfrac{E \times I}{1000}$	$\dfrac{E \times I \times 2}{1000}$	$\dfrac{E \times I \times 1.73}{1000}$
Horsepower	$\dfrac{E \times I \times \%EFF}{746}$	$\dfrac{E \times I \times \%EFF \times PF}{746}$	$\dfrac{E \times I \times \%EFF \times PF \times 2}{746}$	$\dfrac{E \times I \times \%EFF \times PF \times 1.73}{746}$

Percent Efficiency = %EFF = $\dfrac{\text{Output (Watts)}}{\text{Input (Watts)}}$ Power Factor = PF = $\dfrac{\text{Power Used (Watts)}}{\text{Apparent Power}}$ = $\dfrac{KW}{KVA}$

$$E = Volts$$
$$I = Amperes$$
$$W = Watts$$

Note: Direct-current formulas do not use (PF, 2 or 1.73).
Single-phase formulas do not use (2 or 1.73).
Two-phase/four-wire formulas do not use (1.73).
Three-phase formulas do not use (2).

169

U.S. WEIGHTS AND MEASURES

Linear Measures

	1 Inch	= 2.540 Centimeters
12 Inches	=1 Foot	= 3.048 Decimeters
3 Feet	=1 Yard	= 9.144 Decimeters
5.5 Yards	=1 Rod	= 5.029 Meters
40 Rods	=1 Furlong	= 2.018 Hectometers
8 Furlongs	=1 Mile	= 1.609 Kilometers

Mile Measurements

1 Statute Mile	=	5280	Feet
1 Scots Mile	=	5952	Feet
1 Irish Mile	=	6720	Feet
1 Russian Verst	=	3504	Feet
1 Italian Mile	=	4401	Feet
1 Spanish Mile	=	15084	Feet

Other Linear Measurements

1 Hand = 4 Inches	1 Link = 7.92 Inches
1 Span = 9 Inches	1 Fathom = 6 Feet
1 Chain = 22 Yards	1 Furlong = 10 Chains
	1 Cable = 608 Feet

Square Measures

144	Square Inches	= 1	Square Foot
9	Square Feet	= 1	Square Yard
30¼	Square Yards	= 1	Square Rod
40	Rods	= 1	Rood
4	Roods	= 1	Acre
640	Acres	= 1	Square Mile
1	Square Mile	= 1	Section
36	Sections	= 1	Township

Cubic or Solid Measures

1 Cubic Foot	=	1728	Cubic Inches
1 Cubic Yard	=	27	Cubic Feet
1 Cubic Foot	=	7.48	Gallons
1 Gallon (Water)	=	8.34	Pounds
1 Gallon (U.S.)	=	231	Cubic Inches of Water
1 Gallon (IMperial)	=	277¼	Cubic Inches of Water

🔧 U.S. WEIGHTS AND MEASURES

Liquid Measurements

1 Pint	=	4	Gills
1 Quart	=	2	Pints
1 Gallon	=	4	Quarts
1 Firkin	=	9	Gallons (Ale or Beer)
1 Barrel	=	42	Gallons (Petroleum or Crude Oil)

Dry Measure

1 Quart	=	2	Pints
1 Peck	=	8	Quarts
1 Bushel	=	4	Pecks

Weight Measurement (Mass)

A. Avoirdupois Weight

1 Ounce	=	16	Drams
1 Pound	=	16	Ounces
1 Hundredweight	=	100	Pounds
1 Ton	=	2000	Pounds

B. Troy Weight

1 Carat	=	3.17	Grains
1 Pennyweight	=	20	Grains
1 Ounce	=	20	Pennyweights
1 Pound	=	12	Ounces
1 Long Hundred-Weight	=	112	Pounds
1 Long Ton	=	20	Long Hundredweights
	=	2240	Pounds

C. Apothecaries Weight

1 Scruple	= 20	Grains	=	1.296	Grams
1 Dram	= 3	Scruples	=	3.888	Grams
1 Ounce	= 8	Drams	=	31.1035	Grams
1 Pound	= 12	Ounces	=	373.2420	Grams

D. Kitchen Weights and Measures

1 U.S. Pint	=	16	Fluid Ounces
1 Standard Cup	=	8	Fluid Ounces
1 Tablespoon	=	0.5	Fluid Ounces (15 Cubic Centimeters)
1 Teaspoon	=	0.16	Fluid Ounces (5 Cubic Centimeters)

🔲 METRIC SYSTEM

Prefixes

A. Mega =	1000000	E. Deci	=	0.1
B. Kilo =	1000	F. Centi	=	0.01
C. Hecto =	100	G. Milli	=	0.001
D. Deka =	10	H. Micro	=	0.000001

Linear Measure

(The Unit is the Meter = 39.37 Inches)

1 Centimeter	=	10 Millimeters	=	0.3937011	Inch
1 Decimeter	=	10 Centimeters	=	3.9370113	Inches
1 Meter	=	10 Decimeters	=	1.0936143	Yards
			=	3.2808429	Feet
1 Dekameter	=	10 Meters	=	10.936143	Yards
1 Hectometer	=	10 Dekameters	=	109.36143	Yards
1 Kilometer	=	10 Hectometers	=	0.62137	Mile
1 Myriameter	= 10000 Meters				

Square Measure

(The Unit is the Square Meter = 1549.9969 SQ. Inches)

1 SQ. Centimeter	= 100 SQ. Millimeters	=	0.1550	Square Inch
1 SQ. Decimeter	= 100 SQ. Centimeters	=	15.550	Square Inches
1 SQ. Meter	= 100 SQ. Decimeters	=	10.7639	Square Feet
1 SQ. Dekameter	= 100 SQ. Meters	= 119.60		Square Yards
1 SQ. Hectometer	= 100 SQ. Dekameters			
1 SQ. Kilometer	= 100 SQ. Hectometers			

(The Unit is the "Are" = 100 SQ. Meters)

1 Centiare	=	10	Milliares	=	10.7643	Square Feet
1 Deciare	=	10	Centiares	= 11.96033		Square Yards
1 Are	=	10	Deciares	= 119.6033		Square Yards
1 Decare	=	10	Ares	=	0.247110	Acres
1 Hectare	=	10	Decares	=	2.471098	Acres
1 SQ. Kilometer	= 100		Hectares	=	0.38611	Square Mile

Cubic Measure

(The Unit is the "Stere" = 61025.38659 CU. Inches)

1 Decistere	=	10 Centisteres	=	3.531562	Cubic Foot
1 Stere	=	10 Decisteres	=	1.307986	Cubic Yards
1 Dekastere	=	10 Steres	= 13.07986		Cubic Yards

⚓ METRIC SYSTEM

Cubic Measure (continued)

(The Unit is the Meter = 39.37 inches)

1 CU. Centimeter	= 1000 CU. Millimeters	=	0.06102	Cubic Inches
1 CU. Decimeter	= 1000 CU. Centimeters	=	61.02374	Cubic Inches
1 CU. Meter	= 1000 CU. Decimeters	=	35.31467	Cubic Feet
	= 1 Stere	=	1.30795	Cubic Yards
1 CU. Centimeter (Water)		=	1 Gram	
1000 CU. Centimeters (Water) = 1 Liter		=	1 Kilogram	
1 CU. Meter (1000 Liters)		=	1 Metric Ton	

Measures of Weight

(The Unit is the Gram = 0.035274 Ounces)

1 Milligram	=		=	0.015432	Grains
1 Centigram	=	10 Milligrams	=	0.15432	Grains
1 Decigram	=	10 Centigrams	=	1.5432	Grains
1 Gram	=	10 Decigrams	=	15.4323	Grains
1 Dekagram	=	10 Grams	=	5.6438	Drams
1 Hectogram	=	10 Dekagrams	=	3.5274	Ounces
1 Kilogram	=	10 Hectograms	=	2.2046223	Pounds
1 Myriagram	=	10 Kilograms	=	22.046223	Pounds
1 Quintal	=	10 Myriagrams	=	1.986412	Hundredweight
1 Metric Ton	=	10 Quintal	= 22045.622		Pounds
1 Gram	=	0.56438 Drams			
1 Dram	=	1.77186 Grams			
	=	27.3438 Grains			
1 Metric Ton	= 2204.6223 Pounds				

Measures of Capacity

(The Unit is the Liter =1.0567 Liquid Quarts)

1 Centiliter	= 10 Milliliters	=	0.338	Fluid Ounces
1 Deciliter	= 10 Centiliters	=	3.38	Fluid Ounces
1 Liter	= 10 Deciliters	=	33.8	Fluid Ounces
1 Dekaliter	= 10 Liters	=	0.284	Bushel
1 Hectoliter	= 10 Dekaliters	=	2.84	Bushels
1 Kiloliter	= 10 Hectoliters	= 264.2		Gallons

Note: $\dfrac{\text{Kilometers}}{8} \times 5 = \text{Miles}$ or $\dfrac{\text{Miles}}{5} \times 8 = \text{Kilometers}$

METRIC SYSTEM

Metric Designator and Trade Sizes

Metric Designator

12	16	21	27	35	41	53	63	78	91	103	129	155
⅜	½	¾	1	1¼	1½	2	2½	3	3½	4	5	6

Trade Size

Source: NFPA 70, National Electrical Code®, NFPA, Quincy, MA, 2020, Table 300.1 (C), as modified.

U.S. Weights and Measures/Metric Equivalent Chart

	In	Ft	Yd.	Mile	Mm	Cm	M	Km
1 inch =	1	.0833	.0278	1.578×10^5	25.4	2.54	.0254	2.54×10^3
1 Foot =	12	1	.333	1.894×10^4	304.8	30.48	.3048	3.048×10^4
1 Yard =	36	3	1	5.6818×10^4	914.4	91.44	.9144	9.144×10^4
1 Mile =	63360	5280	1760	1	1609344	160934.4	1609.344	1609344
1 mm =	.03937	.00032808	1.0036×10^3	6.2137×10^7	1	0.1	0.001	0.000001
1 cm =	.3937	.0328084	.0109361	6.2137×10^6	10	1	0.01	0.00001
1 m =	39.37	3.280.84	1.093.61	6.2137×10^4	1000	100	1	0.001
1 km =	39370	3280.84	1093.61	0.62137	1000000	100000	1000	1

In = Inches Ft = Foot Yd. = Yard Mm = Millimeter Cm = Centimeter M = Meter Km = Kilometer

Explanation of Scientific Notation

Scientific notation (powers of 10) is simply a way of expressing very large or very small numbers in a more compact format. Any number can be expressed as a number between 1 and 10, multiplied by a power of 10 (which indicates the correct position of the decimal point in the original number). Numbers greater than 10 have positive powers of 10, and numbers less than 1 have negative powers of 10.

Example: $186000 = 1.86 \times 10^5$ $0.000524 = 5.24 \times 10^4$

Useful Conversions/Equivalents

1 BTU	Raises 1 lb of water 1°F
1 Gram Calorie	Raises 1 gram of water 1°C
1 Circular Mil	Equals 0.7854 sq. mil
1 SQ. Mil	Equals 1.2732 cir. mils
1 Mil	Equals 0.001 in.

To determine circular mil (cmil) of a conductor:

Round Conductor . . . . cmil = (Diameter in mils)^2

$$\text{Bus Bar} \qquad \text{cmil} = \frac{\text{Width (mils) x Thickness (mils)}}{0.7854}$$

Notes: 1 millimeter = 39.37 mils   1 cir. millimeter = 1550 cir. mils 1 sq. millimeter = 1974 cir. mils

🔌 TWO-WAY CONVERSION TABLE

To convert from the unit of measure in Column B to the unit of measure in Column C, multiply the number of units in Column B by the multiplier in Column A. To convert from Column C to B, use the multiplier in Column D.

Example: To convert 1000 BTUs to Calories, find the "BTU - Calorie" combination in Columns B and C. "BTU" is in Column B and "Calorie" is in Column C, so we are converting from B to C. Therefore, we use the Column A multiplier. 1000 BTUs x 251.996 = 251996 Calories.

To convert 251996 Calories to BTUs, use the same "BTU - Calorie" combination. But this time you are converting from C to B. Therefore, use the Column D multiplier. 251996 Calories x 0.0039683 = 1000 BTUs.

$$A \times B = C \qquad \& \qquad D \times C = B$$

	To Convert from B to C, Multiply B x A.		To Convert from C to B, Multiply C x D:
A	**B**	**C**	**D**
43560	Acre	Sq. foot	2.2956×10^{-5}
1.5625×10^{-3}	Acre	Sq. mile	640
6,4516	Ampere per sq. cm.	Ampere per sq. in.	0.155003
1.256637	Ampere (turn)	Gilberts	0.79578
33.89854	Atmosphere	Foot of H₂0	0.029499
29.92125	Atmosphere	Inch of Hg	0.033421
14.69595	Atmosphere	Pound force/sq. In.	0.06804
251.996	BTU	Calorie	3.96832×10^{-3}
778.169	BTU	Foot-pound force	1.28507×10^{-3}
$3.93\,10 \times 10^{-4}$	BTU	Horsepower-hour	2544.43
1055.056	BTU	Joule	9.47817×10^{-4}
2.9307×10^{-4}	BTU	Kilowatt-hour	3412.14
3.93015×10^{-4}	BTU/hour	Horsepower	2544.43
$2.93\,01 \times 10^{-4}$	BTU/hour	Kilowatt	3412.1412
0.293071	BTU/hour	Wan	3.41214
4.19993	BTU/minute	Calorle/second	0.23809
0.0235809	BTU/minute	Horsepower	42.4072
17.5843	BTU/minute	Walt	0.0568

(continued on next page)

To Convert from B to C, Multiply B × A.			To Convert from C to B, Multiply C × D:
A	**B**	**C**	**D**
4.1868	Calorie	Joule	0.238846
0.0328084	Centimeter	Foot	30.48
0.3937	Centimeter	Inch	2.54
0.00001	Centimeter	Kilometer	100000
0.01	Centimeter	Meter	100
6.2137×10^{-6}	Centimeter	Mile	160934.4
10	Centimeter	Millimeter	0.1
0.010936	Centimeter	Yard	91.44
7.85398×10^{-7}	Circular mil	Sq. inch	1.273239×10^{6}
0.000507	Circular mil	Sq. millimeter	1973.525
0.06102374	Cubic centimeter	Cubic inch	16.387065
0.028317	Cubic foot	Cubic meter	35.31467
1.0197×10^{-3}	Dyne	Gram force	980.665
1×10^{-5}	Dyne	Newton	100000
1	Dyne centimeter	Erg	1
7.376×10^{-8}	Erg	Foot pound force	1.355818×10^{7}
2.777×10^{-14}	Erg	Kilowatt hour	3.6×10^{13}
1.0×10^{-7}	Erg/second	Watt	1.0×10^{7}
12	Foot	Inch	0.0833
3.048×10^{-4}	Foot	Kilometer	3280.84
0.3048	Foot	Meter	3.28084
1.894×10^{-4}	Foot	Mile	5280
304.8	Foot	Millimeter	0.00328
0.333	Foot	Yard	3
10.7639	Foot candle	Lux	0.0929
0.882671	Foot of H,0	Inch of Hg	1.13292
5.0505×10^{-7}	Foot pound force	Horsepower hour	1.98×10^{6}
1.35582	Foot pound force	Joule	0.737562
3.76616×10^{-7}	Foot pound force	Kilowatt hour	2.655223×10^{6}
3.76616×10^{-4}	Foot pound force	Watt hour	2655.22
3.76616×10^{-7}	Foot pound force/hour	Kilowatt	2.6552×10^{6}
3.0303×10^{-5}	Foot pound force/minute	Horsepower	33000

(continued on next page)

🔁 TWO-WAY CONVERSION TABLE

To Convert from B to C, Multiply B × A.			To Convert from C to B, Multiply C × D:
A	**B**	**C**	**D**
2.2597×10^{-5}	Foot pnd. force/minute	Kilowatt	44253.7
0.022597	Foot pnd. force/minute	Watt	44.2537
1.81818×10^{-3}	Foot pnd. force/second	Horsepower	550
1.355818×10^{-3}	Foot pnd. force/second	Kilowatt	737.562
0.7457	Horsepower	Kilowatt	1.34102
745.7	Horsepower	Watt	0.00134
0.0022046	Gram	Pound mass	453.592
2.54×10^{-5}	Inch	Kilometer	39370
0.0254	Inch	Meter	39.37
1.578×10^{-5}	Inch	Mile	63360
25.4	Inch	Millimeter	0.03937
0.0278	Inch	Yard	36
0.07355	Inch of H_2O	Inch of Hg	13.5951
2.7777×10^{-7}	Joule	Kilowatt hour	3.6×10^{6}
2.7777×10^{-4}	Joule	Watt hour	3600
1	Joule	Watt second	1
1000	Kilometer	Meter	0.001
0.62137	Kilometer	Mile	1.609344
1000000	Kilometer	Millimeter	0.000001
1093.61	Kilometer	Yard	9.144×10^{-4}
0.000621	Meter	Mile	1609.344
1000	Meter	Millimeter	0.001
1.0936	Meter	Yard	0.9144
1609344	Mile	Millimeter	6.2137×10^{-7}
1760	Mile	Yard	5.681×10^{-4}
1.0936×10^{-3}	Millimeter	Yard	914.4
0.224809	Newton	Pound force	4.44822
0.03108	Pound	Slug	32.174
0.0005	Pound	Ton (short)	2000
0.155	Sq. centimeter	Sq. inch	6.4516
0.092903	Sq. foot	Sq. meter	10.76391
0.386102	Sq. kilometer	Sq. mile	2.589988

CENTIGRADE AND FAHRENHEIT THERMOMETER SCALES

°C	°F	°C	°F	°C	°F	°C	°F
0	32						
1	33.8	26	78.8	51	123.8	76	168.8
2	35.6	27	80.6	52	125.6	77	170.6
3	37.4	28	82.4	53	127.4	78	172.4
4	39.2	29	84.2	54	129.2	79	174.2
5	41	30	86	55	131	80	176
6	42.8	31	87.8	56	132.8	81	177.8
7	44.6	32	89.6	57	134.6	82	179.6
8	46.4	33	91.4	58	136.4	83	181.4
9	48.2	34	93.2	59	138.2	84	183.2
10	50	35	95	60	140	85	185
11	51.8	36	96.8	61	141.8	86	186.8
12	53.6	37	98.6	62	143.6	87	188.6
13	55.4	38	100.4	63	145.4	88	190.4
14	57.2	39	102.2	64	147.2	89	192.2
15	59	40	104	65	149	90	194
16	60.8	41	105.8	66	150.8	91	195.8
17	62.6	42	107.6	67	152.6	92	197.6
18	64.4	43	109.4	68	154.4	93	199.4
19	66.2	44	111.2	69	156.2	94	201.2
20	68	45	113	70	158	95	203
21	69.8	46	114.8	71	159.8	96	204.8
22	71.6	47	116.6	72	161.6	97	206.6
23	73.4	48	118.4	73	163.4	98	208.4
24	75.2	49	120.2	74	165.2	99	210.2
25	77	50	122	75	167	100	212

1. Temp. °C = $\frac{5}{9}$ x (Temp. °F – 32)
2. Temp. °F = ($\frac{9}{5}$ x Temp. °C) + 32
3. Ambient temperature is the temperature of the surrounding cooling medium.
4. Rated temperature rise is the permissible rise in temperature above ambient when operating under load.

🔌 FRACTIONS

Definitions

A. A <u>fraction</u> is a quantity less than a unit.

B. A <u>numerator</u> is the term of a fraction indicating how many of the parts of a unit are to be taken. In a common fraction, it appears above or to the left of the line.

C. A <u>denominator</u> is the term of a fraction indicating the number of equal parts into which the unit is divided. In a common fraction, it appears below or to the right of the line.

D. *Examples:*

$$(1) \quad \frac{1}{2} \begin{array}{l} \longrightarrow \text{Numerator} \\ \longrightarrow \text{Denominator} \end{array} = \text{Fraction}$$

$$(2) \quad \text{Numerator} \longrightarrow \text{½} \longleftarrow \text{Denominator}$$

To Add or Subtract

To solve: $\frac{1}{2} - \frac{2}{3} + \frac{3}{4} - \frac{5}{6} + \frac{7}{12} = ?$

A. Determine the lowest common denominator that each of the denominators 2, 3, 4, 6, and 12 will divide into an even number of times.

The lowest common denominator is 12.

B. Work one fraction at a time using the formul:

$$\frac{\text{Common Denominator}}{\text{Denominator of Fraction}} \quad \times \quad \textbf{Numerator of Fraction}$$

(1.) $\frac{12}{2} \times 1 = 6 \times 1 = 6$ ½ becomes $\frac{6}{12}$

(2.) $\frac{12}{3} \times 2 = 4 \times 2 = 8$ ⅔ becomes $\frac{8}{12}$

(3.) $\frac{12}{4} \times 3 = 3 \times 3 = 9$ ¾ becomes $\frac{9}{12}$

(4.) $\frac{12}{6} \times 5 = 2 \times 5 = 10$ ⅚ becomes $\frac{10}{12}$

(5.) $\frac{7}{12}$ remains $\frac{7}{12}$

To Add or Subtract (*continued*)

C. We can now convert the problem from its original form to its new form using 12 as the common denominator.

$^1/_2 - {}^2/_3 + {}^3/_4 - {}^5/_6 + {}^7/_{12}$ = Original form

$\dfrac{6 - 8 + 9 - 10 + 7}{12}$ = Present form

$\dfrac{4}{12} = \dfrac{1}{3}$ Reduced to lowest form

D. To convert fractions to decimal form, simply divide the numerator of the fraction by the denominator of the fraction.

Example: $\dfrac{1}{3}$ = 1 divided by 3 = 0.333

To Multiply

A. The numerator of fraction #1 times the denominator of fraction #2 is equal to the numerator of the quotient.

B. The denominator of fraction #1 times the numerator of fraction #2 is equal to the denominator of the quotient.

C *Example:*

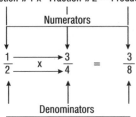

Fraction #1 x Fraction #2 = Product

Numerators

$$\frac{1}{2} \quad \text{x} \quad \frac{3}{4} \quad = \quad \frac{3}{8}$$

Denominators

Note: To change $^3/_8$ to decimal form, divide 3 by 8. This equals 0.375.

🔌 FRACTIONS

To Divide

A. The numerator of fraction #1 times the denominator of fraction #2 is equal to the numerator of the quotient.

B. The denominator of fraction #1 times the numerator of fraction #2 is equal to the denominator of the quotient

C. **Example:** $\dfrac{1}{2} \div \dfrac{3}{4}$

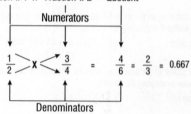

Fraction #1 x Fraction #2 = Quotient

Numerators

$$\dfrac{1}{2} \;\times\; \dfrac{3}{4} \;=\; \dfrac{4}{6} \;=\; \dfrac{2}{3} \;=\; 0.667$$

Denominators

D. An alternate method for dividing by a fraction is to multiply by the reciprocal of the divisor (the second fraction in a division problem).

E. **Example:** $\dfrac{1}{2} \div \dfrac{3}{4}$

The reciprocal of $\dfrac{3}{4}$ is $\dfrac{4}{3}$

So, $\dfrac{1}{2} + \dfrac{3}{4} \;=\; \dfrac{1}{2} \;\times\; \dfrac{4}{3} \;=\; \dfrac{4}{6} \;=\; \dfrac{2}{3} \;=\; 0.667$

🔌 EQUATIONS

The word "Equation" means equal or the same as.

Example:
$$2 \times 10 = 4 \times 5$$
$$20 = 20$$

Rules

A. **The same number may be added to both sides of an equation without changing its values.**

Example:
$$(2 \times 10) + 3 = (4 \times 5) + 3$$
$$23 = 23$$

B. **The same number may be subtracted from both sides of an equation without changing its values.**

Example:
$$(2 \times 10) - 3 = (4 \times 5) - 3$$
$$17 = 17$$

C. **Both sides of an equation may be divided by the same number without changing its values.**

Example:
$$\frac{2 \times 10}{20} = \frac{4 \times 5}{20}$$
$$1 = 1$$

D. **Both sides of an equation may be multiplied by the same number without changing its values.**

Example:
$$3 \times (2 \times 10) = 3 \times (4 \times 5)$$
$$60 = 60$$

E. **Transposition:**

The process of moving a quantity from one side of an equation to the other side of an equation by changing its sign of operation.

1. A term may be transposed if its sign is changed from plus (+) to minus (–), or from minus (–) to plus (+).

Example:
$$X + 5 = 25$$
$$X + 5 - 5 = 25 - 5$$
$$X = 20$$

EQUATIONS

E. Transposition (*continued*):

2. A multiplier may be removed from one side of an equation by making it a divisor on the other side; or a divisor may be removed from one side of an equation by making it a multiplier on the other side.

Example: Multiplier from one side of equation (4) becomes divisor on other side.

$$4X = 40 \text{ becomes } X = \frac{40}{4} = 10$$

Example: Divisor from one side of equation becomes multiplier on other side.

$$\frac{X}{4} = 10 \text{ becomes } X = 10 \times 4$$

Signs

A. Addition of numbers with *different* signs:

1. **Rule: Use the sign of the larger and subtract.**

Example:

$$\begin{array}{r} +3 \\ +-2 \\ \hline +1 \end{array} \qquad \begin{array}{r} +2 \\ ++3 \\ \hline +1 \end{array}$$

B. Addition of numbers with the *same* signs:

2. **Rule: Use the common sign and add.**

Example:

$$\begin{array}{r} +3 \\ ++2 \\ \hline +5 \end{array} \qquad \begin{array}{r} -3 \\ +-2 \\ \hline -5 \end{array}$$

C. Subtraction of numbers with *different* signs:

3. **Rule: Change the sign of the subtrahend (the second number in a subtraction problem) and proceed as in addition.**

Example:

$$\begin{array}{r} +3 \\ --2 \\ \hline \end{array} = \begin{array}{r} +3 \\ ++2 \\ \hline +5 \end{array} \qquad \begin{array}{r} -2 \\ -+3 \\ \hline \end{array} = \begin{array}{r} -2 \\ +-3 \\ \hline -5 \end{array}$$

⊓ EQUATIONS

Signs (continued)

D. **Subtraction** of numbers with the *same* signs:

    4. **Rule: Change the sign of the subtrahend (the second number in a subtraction problem) and proceed as in addition**

Example:

$$\begin{array}{c} +3 \\ -+2 \end{array} = \begin{array}{c} +3 \\ +\;-2 \\ \hline +1 \end{array} \qquad \begin{array}{c} -3 \\ -\;-2 \end{array} = \begin{array}{c} -3 \\ +\;+2 \\ \hline -1 \end{array}$$

E. **Multiplication:**

    5. **Rule: The product of any two numbers having <u>like</u> signs is <u>positive</u>. The product of any two numbers having <u>unlike</u> signs is <u>nenative</u>.**

Example:
$$(+3) \times (-2) = -6$$
$$(+3) \times (-2) = -6$$
$$(+3) \times (-2) = -6$$
$$(+3) \times (-2) = -6$$

F. **Division:**

    6. **Rule: If the divisor and the dividend have <u>like</u> signs, the sign of the quotient is <u>positive.</u> If the divisor and dividend have <u>unlike</u> signs, the sign of the quotient is <u>negative.</u>**

Example:
$$\frac{+6}{-2} = -3 \qquad\qquad \frac{+6}{+2} = +3$$

$$\frac{-6}{+2} = -3 \qquad\qquad \frac{-6}{-2} = +3$$

⚡ ELECTRICAL SAFETY DEFINITIONS

Note: Some NFPA 70E definitions include informational notes, which are shown below. Comments shown in italics under some definitions are additional explanations that do not appear in NFPA 70E.

Arc-Flash Hazard: A source of possible injury or damage to health associated with the possible release of energy caused by an electric arc.

Informational Note No. 1: The likelihood of occurrence of an arc flash incident increases when energized electrical conductors or circuit parts are exposed or when they are within equipment in a guarded or enclosed condition, provided a person is interacting with the equipment in such a manner that could cause an electric arc. An arc flash incident is not likely to occur under normal operating conditions when enclosed energized equipment has been properly installed and maintained.

Boundary, Arc Flash: When an arc flash hazard exists, an approach limit from an arc source at which incident energy equals 1.2 cal/cm² (5 J/cm²).

De-energized: Free from any electrical connection to a source of potential difference and from electrical charge; not having a potential different from that of the earth.

Comment: This is a key concept of NFPA 70E. The safest way to work on electrical conductors and equipment is de-energized. See Electrically Safe Work Condition.

Electrical Hazard: A dangerous condition such that contact or equipment failure can result in electrical shock, arc-flash burn, thermal burn, or arc-blast injury.

Electrical Safety: Identifying hazards associated with the use of electrical energy and taking precautions to reduce the risk associated with those hazards.

⌧ ELECTRICAL SAFETY DEFINITIONS

Note: Some NFPA 70E *definitions include informational notes, which are shown below. Comments shown in italics under some definitions are additional explanations that do not appear in* NFPA 70E

Electrically Safe Work Condition: A state in which an electrical conductor or circuit part has been disconnected from energized parts, locked/tagged in accordance with established standards, tested to verify the absence of voltage, and, if necessary, temporarily grounded for personnel protection.

Comment: This is a key concept of NFPA 70E. The safest way to work on electrical conductors and equipment is de-energized. The process of turning off the electricity, verifying that it is off, and ensuring that it stays off while work is performed is called "establishing an electrically safe work condition. " Many people call the process of ensuring that the current is removed "lockout/tagouthowever, lockout/tagout is only one step in the process.

Energized: Electrically connected to, or is, a source of voltage.

Incident Energy: The amount of thermal energy impressed on a surface, a certain distance from the source, generated during an electrical arc event. Incident energy is typically expressed in calories per square centimeter (cal/cm^2).

Qualified Person: One who has demonstrated skills and knowledge related to the construction and operation of electrical equipment and installations and has received safety training to identify the hazards and reduce the associated risk.

Comment: A person can be considered qualified with respect to certain equipment and methods but still be unqualified for others.

Working Distance: The distance between a person's face and chest area and a prospective arc source.

Informational Note: Incident energy increases as the distance from the arc source decreases.

WHO IS RESPONSIBLE FOR ELECTRICAL SAFETY?

NFPA 70E, like OSHA, states that both employers and employees are responsible for preventing injury.

- Employers shall provide safety-related work practices and shall train the employees.
- Employees shall implement the safety-related work practices established.
- Multiple employers often work together on the same construction site or in buildings and similar facilities. Some might be onsite personnel working for the host employer, while others are "outside" personnel such as electrical contractors, mechanical and plumbing contractors, painters, or cleaning crews. Outside personnel working for the host employer are employees of contract employers.
- NFPA 70E requires that when a host employer and contract employer work together within the limited approach boundary or the arc-flash boundary of exposed energized electrical conductors or circuit parts, they must coordinate their safety procedures.
- Where the host employer has knowledge of hazards covered by NFPA 70E that are related to the contract employer's work, there shall be a documented meeting between the host employer and the contract employer.
- Outside contractors often are required to follow the host employer's safety procedures.
- Multiple employers involved in the same project sometimes decide to follow the most stringent set of safety procedures.
- Whichever approach is taken, the decision should be recorded in the safety meeting documentation. In accordance with NFPA 70E, where the host employer has knowledge of hazards covered by 70E that are related to the contract employer's work, there shall be a documented meeting between the host employer and the contract employer.

🔧 LOCKOUT–TAGOUT AND ELECTRICALLY SAFE WORK CONDITION

The term lockout/tagout refers to specific practices and procedures to safeguard employees from the unexpected energization or startup of machinery and equipment, or the release of hazardous energy during service or maintenance activities. OSHA and NFPA 70E address the control of hazardous energy during service or maintenance of machines or equipment.

OSHA's standard for The Control of Hazardous Energy (Lockout/Tagout), found in Title 29 of the Code of Federal Regulations (CFR) Part 1910.147, addresses the practices and procedures necessary to disable machinery or equipment, thereby preventing the release of hazardous energy while employees perform servicing and maintenance activities. Other OSHA standards, such as 29 CFR 1910.269 and 1910.333 also contain energy control provisions.

Article 120 in NFPA 70E contains requirements for lockout/tagout, as well as procedures for establishing and verifying an electrically safe work condition.

Establishing and.verifying an electrically safe work condition shall include all of the following steps, which shall be performed in the order presented, if feasible:

1. Identify the power sources.

2. Disconnect power sources.

3. If possible, visually verify that power is disconnected.

4. Release stored electrical energy.

5. Release or block stored mechanical energy.

6. Apply lockout/tagout devices.

7. Test for the absence of voltage.

8. Install temporary protective grounding equipment if there is a possibility of induced voltages or stored electrical energy.

INFORMATION USUALLY FOUND ON AN ARC-FLASH EQUIPMENT LABEL

(Courtesy of Charles R. Miller)

Electrical equipment such as switchboards, panelboards, industrial control panels, meter socket enclosures, and motor control centers that are in other than dwelling units, and are likely to require examination, adjustment, servicing, or maintenance while energized, shall be field marked with a label containing the information in NFPA70E, 130.5(H).

The available incident energy at the working distance. Instead of the available incident energy and the corresponding working distance, the arc-flash PPE category could have been on this label. See NFPA70E, 130.5(H).

When an arc flash hazard exists, this is the distance from an arc source at which incident energy equals 1.2 cal/cm². The onset of a second-degree burn is assumed to be when the skin receives 1.2 cal/cm² of incident energy

When incident energy is on the label, it is based on a working so the working distance has to be on the label as well.

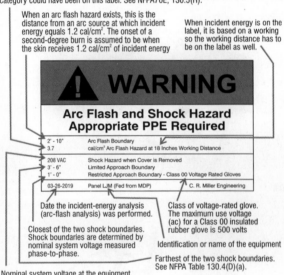

⚠ WARNING

Arc Flash and Shock Hazard
Appropriate PPE Required

2' - 10"	Arc Flash Boundary
3.7	cal/cm² Arc Flash Hazard at 18 Inches Working Distance
208 VAC	Shock Hazard when Cover is Removed
3' - 6"	Limited Approach Boundary
1' - 0"	Restricted Approach Boundary - Class 00 Voltage Rated Gloves
03-26-2019	Panel LJM (Fed from MDP) C. R. Miller Engineering

Date the incident-energy analysis (arc-flash analysis) was performed.

Class of voltage-rated glove. The maximum use voltage (ac) for a Class 00 insulated rubber glove is 500 volts.

Closest of the two shock boundaries. Shock boundaries are determined by nominal system voltage measured phase-to-phase.

Identification or name of the equipment

Farthest of the two shock boundaries. See NFPA Table 130.4(D)(a).

Nominal system voltage at the equipment.

🔧 SELECTING AND USING TEST INSTRUMENTS

Single Instrument

A multimeter combines the voltmeter, ohmmeter, and milliammeter into a single instrument. In the field, this one instrument can be used to measure (alternating current) ac and (direct current) dc voltages, ac and dc current flow, and electrical resistance.

Although analog multimeters have long been available, most multimeters used today are digital multimeters (DMMs).

Analog:

Digital:

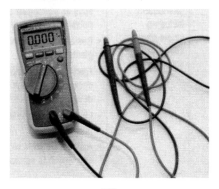

SELECTING AND USING TEST INSTRUMENTS

Selecting an Appropriate Multimeter

- Choose a meter that is properly rated for the circuit or component to be tested.

- Ensure that both the meter and the leads have the correct category rating for each task.

- Check that the ratings of the leads or accessories meet or exceed the rating of the multimeter.

- Ensure that the meter case is not wet, oily, or cracked; that the input jacks are not broken; and that there are no other obvious signs of damage.

- Check the test leads carefully. Ensure that the insulation is not cut, cracked, or melted and that the tips are not loose.

Testing a Multimeter

To measure continuity:

- Set the multimeter to the lowest setting for resistance.

- Touch the tips of the two probes together.

- The display should show 0 ohms (Ω). A DMM will typically display OL (overload or out of limits) when a circuit is open (lacks continuity).

- If OL is displayed during this test, either the meter is defective or the test leads are defective.

To check a meter's ability to measure resistance:

- Use a resistor with a known value.

- Set the multimeter to measure resistance.

- Place one probe at each end of the resistor.

🔌 SELECTING AND USING TEST INSTRUMENTS

- The value in the display should be very close to the known value of the resistor being used for the test.

- If the multimeter does not measure the resistance correctly, double-check it by repeating the test with a different resistor of a known value.

To check the dc voltage function:

- Use a new battery.

- Set the multimeter to measure dc voltage.

- Place the red probe on the positive terminal of the battery and the black probe on the negative terminal.

- The multimeter should display a reading of, or close to, the battery's labeled voltage.

- Reversing the leads on a digital meter will display a negative (–) reading.

To check the ac-voltage function:

- Use a 120 VAC receptacle known to be energized.

- Plug in and turn on a lamp or another simple device to ensure that the receptacle is energized. Set the multimeter to measure ac voltage.

- Place the red probe into the energized side—the smaller slot of the receptacle—and the black probe into the other side (neutral).

- The multimeter should display a reading of, or close to, 120 VAC.

If the multimeter fails any one of these tests, it is best to return it to the manufacturer for service. Technicians should never attempt to open and repair the meter's electronics or movement. If the multimeter will not perform any functions or provide a display when tested, the problem may be weak or dead batteries. DMMs have a battery indicator that will display when the battery is weak but still functional. Always replace the battery as soon as the indication is displayed to ensure continued reliable operation.

If the batteries are not the problem, determine if there is an accessible fuse inside the battery compartment. If there is, check to see whether it is open; another meter may be needed to check the continuity of the fuse if its condition cannot be determined visually. Replace an open fuse with one of the proper current and voltage rating. Using the wrong fuse, especially one that has a higher current rating than it should, can result in the meter being damaged beyond repair.

Return a multimeter to the manufacturer for an annual calibration to ensure its accuracy.

Common Testing Errors

Try to avoid the following common errors when using a multimeter to measure voltage:

- Measuring voltage while the red test lead is in the wrong jack. The test-lead jack for the red lead is usually marked with the V symbol and is often marked with the Ω symbol. Applying voltage to the meter while the red lead is plugged into the wrong jack may damage the meter.

- Measuring ac voltage on a dc setting. The dial symbol usually associated with measuring ac voltage is a V with a wavy line above it, or an mV with the same wavy line above. The latter is for measuring in the millivolt range.

- Keeping the test probe in contact with an energized surface longer than necessary. The longer it is in contact, the more time there is for an accident to occur.

- Using the meter above its rated voltage. Technicians should have a general idea of the expected voltage or current to be measured and ensure that the meter is capable of safely making such readings.

![icon] SELECTING AND USING TEST INSTRUMENTS

Using a Multimeter to Measure Voltage

Take all available and necessary precautions to avoid contact with energized surfaces, and ensure that the probe of the test lead does not accidentally bridge two points with different electrical potentials. Use test leads with finger guards, and always keep fingers behind them.

- Select an appropriate multimeter for the job,

- Visually inspect it and the test leads for any signs of damage. Before using the multimeter on the job, always test it on a known voltage source to verify it is functioning properly.

- Insert the black test lead into the common input jack (COM) and the red test lead into the input jack for ac volts (V or VΩ). Take care to insert the test leads into the correct jacks.

- Select ac as the type of voltage to be measured. If the meter has no auto-ranging feature, select the voltage range.

- Place the red probe onto the energized side of the circuit.

- Place the black probe onto the neutral side of the circuit or to ground.

- Read the voltage displayed on the meter while both probes make good electrical contact with their individual targets.

Using a Multimeter to Measure Resistance

- The ohmmeter function can be used to determine whether a circuit has continuity or to determine a specific resistance value.

- Select an appropriate multimeter for the job. Visually inspect it and the test leads for any signs of damage.

- Test the multimeter to verify it is functioning properly and can measure continuity and resistance. A small resistor of a known value can be used to verify meter operation.

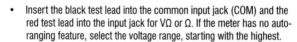

SELECTING AND USING TEST INSTRUMENTS

- Insert the black test lead into the common input jack (COM) and the red test lead into the input jack for VΩ or Ω. If the meter has no auto-ranging feature, select the voltage range, starting with the highest.

- Set the meter to measure resistance (Ω).

- Ensure the circuit is deenergized.

- Touch the test leads to the two points in the circuit across the resistance.

- Read and record the resistance value displayed on the meter.

Using a Multimeter to Measure Current

A technician will frequently use a meter to measure ac. In some cases, ac values in the milliamp, or even the microamp (μA), range may need to be measured.

Most multimeters can measure these small current values by placing the meter in series with the circuit and then energizing the circuit to read the current flow.

Multimeters typically have a maximum current limit of 10 A when using this method of measurement. Do not exceed the maximum current limit of the meter.

Remember that exceeding the current limit with the meter in series with the load can result in significant damage to the meter.

To measure current values that exceed the meter's limit, use either a clamp-on ammeter or a clamp-on ammeter accessory designed for the multimeter in use.

Selecting a Clamp-On Ammeter

Many clamp-on ammeters are designed to provide most or all the functions of a DMM, with the added convenience of a built-in ammeter that can measure a significant amount of current. Therefore, they may be used to measure voltage, resistance, and current like any DMM. A major distinguishing feature of a clamp-on ammeter is that it has a current transformer built into the jaws that can be opened and closed around a conductor. Closing the jaws around a conductor enables the current flow to be measured without having to handle the conductor or disrupt the operation of the component or system being tested. Through induction, a small current is induced in the jaws by the current in the conductor.

Typical clamp-on ammeter features and functions include, but are not limited to, the following:

- A selector switch for selecting the desired test function (voltage, current, or resistance)

- An auto-ranging feature to automatically select the proper measurement range

- A HOLD function to freeze the reading shown on the display

- A minimum/maximum memory function to determine the highest and lowest reading over the course of a test. (The maximum function is valuable for measuring the inrush current when a motor starts.)

- A capacitor-testing feature that measures capacitance and checks for shorts and opens

- A continuity beeper that is activated when continuity through a circuit is detected. (The beeper is an advantage fortroubleshooting, particularly in tight spaces, because attention can remain focused on the meter leads—listening for the beeper instead of diverting attention from the leads to check the readings.)

- Overload protection to prevent damage to the meter and to protect the user

Always read and follow the manufacturer's instructions for the meter to make the best use of its features

🔌 SELECTING AND USING TEST INSTRUMENTS

Carefully review an ammeter's specifications, features, and functions. Be sure it has overload protection to protect the meter and the user. Choose an ammeter that can provide the degree of accuracy required. Ensure the ammeter is properly rated for the expected current. Inductive loads, such as electric motors, experience a significant inrush of current when they start. This surge occurs when a motor is first energized and must be brought up to speed from a complete stop. Although the surge may last less than one second, the current can be many times higher than the current shown on the motor data plate, especially if the motor has mechanical or electrical defects. Remember to consider the possible inrush current when selecting an ammeter and the range. The range can be changed, if needed, to a more appropriate level while the ammeter is actively measuring current. Meters with an auto-ranging feature eliminate this concern. Ensure that the ammeter has the correct category rating for each task. Also, be sure that the ratings of the leads or other accessories meet or exceed the rating of the ammeter. When making a current measurement, ensure that the clamp is firmly and completely closed to get an accurate reading.

⚡ SELECTING AND USING TEST INSTRUMENTS

Testing a Clamp-On Ammeter

- Verify that the clamp-on ammeter is working properly before using it. The procedures for testing a clamp-on ammeter are the same as those used to test the functions of a multimeter.

- Always inspect a clamp-on ammeter before using it. Ensure that the case is not cracked or greasy and that there are no obvious signs of damage. Be sure that the jaw tips are not dirty and that they meet and interlock properly. Proper jaw alignment at the tips is essential for an accurate measurement. If the jaws are dirty or misaligned, the meter will not read correctly.

- Test a clamp-on ammeter to verify it can measure continuity/resistance and voltage. Do not use the meter if it fails any one of the tests. If the ammeter will not perform any functions at all when it is tested, check for a blown fuse and/or replace the battery.

Using a Clamp-On Ammeter to Measure Current

Most digital clamp-on ammeters have an auto-ranging feature. If, however, the range needs to be set manually, always start at the highest range and adjust the range as needed to obtain an accurate reading. Place the jaws of the ammeter around only one conductor at a time. Placing the jaws around two conductors at the same time will produce an inaccurate reading.

- While the system is deenergized, select the location for the current measurement and separate the target conductor from others so the jaws can safely and easily be snapped closed around the conductor when the power is on.

- Make any preparations for measurement that can be made while the power is off to significantly reduce exposure to serious injury.

- Note that wire insulation is not a factor; there is no need to place the jaws around bare conductors. In fact, doing so can be dangerous.

🔧 SELECTING AND USING TEST INSTRUMENTS

- If the current reading is expected to be at the low end of the lowest ammeter range, a more accurate reading can be obtained by wrapping the single conductor wire multiple times around one jaw of the ammeter.

- It is not often necessary for digital meters, but it is a significant help to read low current values on an analog model. Each time the wire passes through the jaws, the current it carries is sensed by the meter.

- If the same conductor passes through the jaws five times, the current measured will be five times the actual current flow. Be sure, however, to divide the reading by the number of times the wire passes through the jaws.

- While wrapping the conductor, do not allow the wraps to cross over each other. This can affect the accuracy.

- Note that this works equally well with a clamp-on ammeter accessory used with multimeters.

Non-contact Voltage Tester

🔌 SELECTING AND USING TEST INSTRUMENTS

Although a non-contact voltage tester is not suitable for work that requires any level of accuracy, it is useful for quickly verifying that a circuit is energized or deenergized. It is important to note that it should not replace voltmeter testing to prove that a circuit is deenergized for safety reasons. Electrical safety demands that a more dependable test device be used. The test instrument must be able to test each phase conductor or circuit part both phase-to-phase and phase-to-ground.

Non-contact voltage testers are relatively small, compact, battery-powered instruments that fit in a shirt pocket. They typically have the following features:

- An On/Off power button

- A flashlight or indicator light that shows the tester is operational and the flashlight may help illuminate the workplace (in some models, the flashlight can be operated independently of the voltage tester)

- A high-intensity LED light and/or a warning tone to notify when voltage is detected

- An automatic shutoff to conserve power and extend battery life

- A low-battery indicator

- A built-in self-test feature

Non-contact voltage testers are available in several voltage ranges. A standard model may be sensitive only to voltages above a range of 90-100 VAC.

Dual-voltage models can register and differentiate between standard voltages and lower voltages. This feature makes these models more suitable for testing low-voltage ac circuits for the presence of power.

Note that these tools are not designed to sense the presence of dc voltage, because they rely on the presence of a magnetic field generated by ac. They may occasionally respond to the presence of dc voltage, but not reliably.

SELECTING AND USING TEST INSTRUMENTS

Selecting a Non-contact Voltage Tester

- When selecting a non-contact voltage tester, choose one with a range suitable for the expected voltage.

- Be sure the instrument is sensitive enough in terms of the minimum voltage it can detect. Inspect the tester for any obvious signs of damage. Ensure the batteries for the tester are working.

- Many testers offer a battery test button to ensure the device is working properly.

Testing a Non-contact Voltage Tester

Check the tester on a known, energized voltage source prior to using the instrument in the field. Use the built-in self-test feature, if the tester is so equipped. Otherwise, a simple way to test it is to place the tip of the tester near the line-voltage side (smaller slot) of an electrical outlet known to be energized. If the tester is working properly, the tester should give a clear audible and/or visual indication that voltage is present.

Using a Non-contact Voltage Tester

Although non-contact voltage testers may detect voltages up to 1000 VAC or more, working near voltages this high can be extremely hazardous. They are best used to detect voltages well below that value.

When safety is at stake and the intention is to make contact with conductors and/or electrical parts, use a multimeter to conduct the test to ensure the circuit has been deenergized.

- Select an appropriate non-contact voltage tester for the job.

- Inspect it for any obvious damage.

- Ensure the batteries are working. If the instrument is equipped with a self-test feature, use it to ensure the instrument is functioning properly.

🔌 SELECTING AND USING TEST INSTRUMENTS

- Before using the voltage tester on the job, always test it to verify that it is functioning properly.

- Gradually move the tester as close as possible to the wire you wish to test.

- If voltage is present, the tester will react with both an audio and visual signal.

- If the circuit is off, there should be no voltage present and no response from the tester.

- If the circuit is live, the tester should provide a visual, and possibly an audible, indication that voltage is present.

- Move the tester from one power wire to the other to be sure.

- Cycle the power off and watch for the indication that voltage is no longer present.

Notes

Notes

Notes

Notes

Notes

Notes

Notes

Notes